Software Usability

Edited by Laura M. Castro,
David Cabrero and Rüdiger Heimgärtner

Published in London, United Kingdom

IntechOpen

Supporting open minds since 2005

Software Usability
http://dx.doi.org/10.5772/intechopen.91112
Edited by Laura M. Castro, David Cabrero and Rüdiger Heimgärtner

Contributors
Suraj Singh Senjam, Nelly Condori-Fernandez, Marcela Quispe, Alejandro Catala, Joao Araujo, Patricia Lago, Nicolas Becu, Osvaldo Gervasi, Damiano Perri, Marco Simonetti, Sergio Tasso, Taimur Khan, Achim Ebert, Afzal Gul, Syed Samad Shakeel, Hamza Masud, Ita Richardson, Bilal Ahmad, Sarah Beecham, Nauman Jalil, Sanghamitra Mohanty, Jide Ebenezer Taiwo Akinsola, Samuel Akinseinde, Olamide Kalesanwo, Moruf Adeagbo, Kayode Oladapo, Oladapo, Ayomikun Awoseyi, Funmilayo Abibat Kasali

First published in London, United Kingdom, 2022 by IntechOpen
IntechOpen is the global imprint of INTECHOPEN LIMITED, registered in England and Wales, registration number: 11086078, 5 Princes Gate Court, London, SW7 2QJ, United Kingdom
Printed in Croatia

British Library Cataloguing-in-Publication Data
A catalogue record for this book is available from the British Library

Additional hard and PDF copies can be obtained from orders@intechopen.com

Software Usability
Edited by Laura M. Castro, David Cabrero and Rüdiger Heimgärtner
p. cm.
Print ISBN 978-1-83968-966-6
Online ISBN 978-1-83968-967-3
eBook (PDF) ISBN 978-1-83968-968-0

For EU product safety concerns:
IN TECH d.o.o., Prolaz Marije Krucifikse Kozulić 3, 51000 Rijeka, Croatia, info@intechopen.com or visit our website at intechopen.com.

Meet the editors

Dr. Laura M. Castro is a professor at the University of A Coruña, Spain, where she has been lecturing for fifteen years. She is currently the studies coordinator for the degree in Software Engineering at the same university and is responsible for several courses on Software Architecture and Software Testing. Her most recent research interests focus on the automatic validation of distributed systems. She has supervised three Ph.D. theses and acted as PI in several European projects. She is also actively involved in dissemination activities, in particular the visibility of women in STEM, as a member of the Association for Computing Machinery's Council on Women in Computing (ACM-W) Europe.

Dr. David Cabrereo is an associate professor at the University of A Coruña, Spain, where he leads the Models and Applications of Distributed Systems (MADS) research group. He is currently responsible for the mandatory course on Human-Computer Interaction for CS undergraduate students. He has been working in distributed systems for more than a decade but has combined this with his interests in accessibility and open source. In 2004, he was awarded the "Eclipse Innovation Grant" by IBM.

Dr. Rüdiger Heimgärtner is the founder and owner of Intercultural User Interface Consulting (IUIC) and the author of the first German-language book on intercultural user interface design (IUID). He is a foundation member of the International Usability and User Experience Qualification Board (UXQB) and teaches the scheme for certified professionals for UX (CPUX). As an assessor for SPICE and human-centered design (HCD) processes, he promotes the process assessment model (PAM) for HCD processes at the International Organization for Standardization (ISO) and coaches how to establish and improve usability engineering processes to support, enhance, and optimize software usability.

Contents

Preface **XI**

Section 1
Human Learning Improvement **1**

Chapter 1 **3**
Learning Mathematics in an Immersive Way
by Damiano Perri, Marco Simonetti, Sergio Tasso and Osvaldo Gervasi

Chapter 2 **19**
Usability of Computerised Gaming Simulation for Experiential Learning
by Nicolas Becu

Section 2
Breaking Down Barriers **35**

Chapter 3 **37**
Smartphones for Vision Rehabilitation: Accessible Features and Apps, Opportunity, Challenges, and Usability Evaluation
by Suraj Singh Senjam

Chapter 4 **55**
Usability Recommendations for Designers of Smartphone Applications for Older Adults: An Empirical Study
by Bilal Ahmad, Ita Richardson and Sarah Beecham

Chapter 5 **77**
Indian Language Compatible Intelligent User Interfaces
by Sanghamitra Mohanty

Section 3
Development Methodologies and Techniques **91**

Chapter 6 **93**
Negative UX-Based Approach for Deriving Sustainability Requirements
by Nelly Condori-Fernandez, Marcela Quispe-Cruz, Alejandro Catala, Joao Araujo and Patricia Lago

Chapter 7 **115**
Implementing Visual Analytics Pipelines with Simulation Data
by Taimur Khan, Syed Samad Shakeel, Afzal Gul, Hamza Masud and Achim Ebert

Chapter 8 **139**
Introduction to Intelligent User Interfaces (IUIs)
by Nauman Jalil

Chapter 9 **155**
Application of Artificial Intelligence in User Interfaces Design for Cyber Security Threat Modeling
by Jide Ebenezer Taiwo Akinsola, Samuel Akinseinde, Olamide Kalesanwo, Moruf Adeagbo, Kayode Oladapo, Ayomikun Awoseyi and Funmilayo Kasali

Preface

Software usability should be central to any software development, especially when it comes to software that is going to be used by people. What, if not the *ease of use* of technology, could be a sign of its purpose and intent to help and assist humans?

We can all agree that software usability is not yet at the fore of every development project. For far too long, technology has come first, solving problems indeed, but far too often *forcing* users to do so in certain ways, often designed without the involvement of said users.

For this book, we set out to put together a collection of high-quality contributions that could help broaden developers' and non-developers' minds alike when it comes to considering software usability. Approachable and easy to understand, this volume presents novel research and experiences in the field of software usability. We hope readers, who may not be software *makers* but who are undoubtedly software *users*, will find the information contained herein accessible and useful.

Drs. Laura M. Castro and David Cabrero
Department of Computer Science and Information Technologies,
University of A Coruña,
A Coruña, Spain

Dr. Rüdiger Heimgärtner
Intercultural User Interface Consulting,
Berlin, Germany

Section 1

Human Learning Improvement

Chapter 1

Learning Mathematics in an Immersive Way

Damiano Perri, Marco Simonetti, Sergio Tasso and Osvaldo Gervasi

Abstract

In this work we introduce a learning system based on Virtual Reality and Augmented Reality for studying analytical-geometric structures that are part of the curriculum in mathematics and physics in high school classes. We believe that an immersive study environment has several advantages with respect to traditional two-dimensional environments (such as a book or the simple screen of a PC or even a tablet), such as the spatial understanding of the concepts exposed, more peripheral awareness and moreover an evident decreasing in the information dispersion phenomenon. This does not mean that our teaching proposal has to substitute the traditional approaches, but it can be seen as a robust tool to support learning. In the first phase of our research we have sought to understand which mathematical objects and which tools could have been used to enhance the teaching of mathematics, in order to demonstrate that the use of Virtual Reality and Augmented Reality techniques significantly improves the level of understanding of the mathematical subject being studied by the students. The system which provides for the integration of two machine levels, hardware and software, was subsequently tested by a representative sample of students who then provided feedback through a questionnaire.

Keywords: immersive learning, virtual reality, augmented reality, Unity3D, blender, usability, accessibility

1. Introduction

Virtual Reality (VR) and Augmented Reality (AR) are technologies that since their inception have sometimes suffered fluctuating fates, sometimes due to the lack of suitable low-cost hardware, sometimes due to the inherent complexity of the technologies adopted. With the advent of mobile technologies, their fates have radically changed and today we have both low-cost hardware and software approaches that make them widely usable in many areas of modern life. Nowadays, we find VR and AR everywhere, in manifold applications: from entertainment [1, 2], teaching [3, 4], tourism [5, 6], manufacturing [7, 8], networking and communications [9, 10], microelectronic and high performances hardware industries [11, 12], e-commerce [13], medicine [14–19].

In this work we are going to focus on the adoption of VR and AR technologies teaching mathematics, analyzing the feasibility of the process and analyzing the usability of the implemented software platform.

During our work, we were guided by a specific goal: the possibility of giving a real and visual form to the abstract objects of mathematics. This represents a further development in the visual representation of mathematical concepts, which in the course of history has been evolving from the primitive use representations of simple counting objects, such as the tally sticks, through the elegant structures of the symbolic algebra of the seventeenth century, to the imposing constructions of mathematical analysis and modern geometry, to get to the current and amazing views of numerical analysis through computer graphics. In this context, we have set out to investigate the possibility of enhancing students' understanding of the concept of link between an algebraic-set structure and its geometric representation on an orthogonal Cartesian space (concept best known with the name of function).

In the first phase of the work, our attention has been directed to a certain number of functions, that are used in senior high-school classes, such as the representation of trajectories in the Cartesian plane and simply surfaces in the three-dimensional space.

The very first proposed functions are as follows:

$y = \cos x$ The **cosine function** is well known to students who currently use it to solve trigonometric, analytical and geometric problems. Moreover, it is also present in all fields of physics in order to model the behavior of several phenomena in mechanics, thermodynamics, optics and electrodynamics, just to limit our analysis to some main cases [20]. In **Figure 1** is shown the VR representation of such function.

$z = e^{-x}$ The **exponential function** is one of the most important functions in mathematical analysis, whose applications span to different fields of knowledge and is well studied in high school. Exponential functions can be used in many contexts, such as the compound interest in finance, to study the evolution of the population growth and the radioactive decay [21]. In **Figure 2** is shown the VR representation of the named function.

$z = \ln(x^2 + y^2)$ This function is an example of **bi-dimensional logarithm**, useful to describe astrophysical objects. In **Figure 3** is shown the VR representation of the named function.

$z = \frac{\sin(x^2+y^2)}{x^2+y^2}$ This function is an example of **bi-dimensional dumped sine**, useful to describe objects in fluid dynamics, electronics and telecommunications. In **Figure 4** is shown the VR representation of the function.

Figure 1.
Cosine wave plot.

Figure 2.
Exponential function plot.

Figure 3.
Bi-dimensional natural logarithm plot.

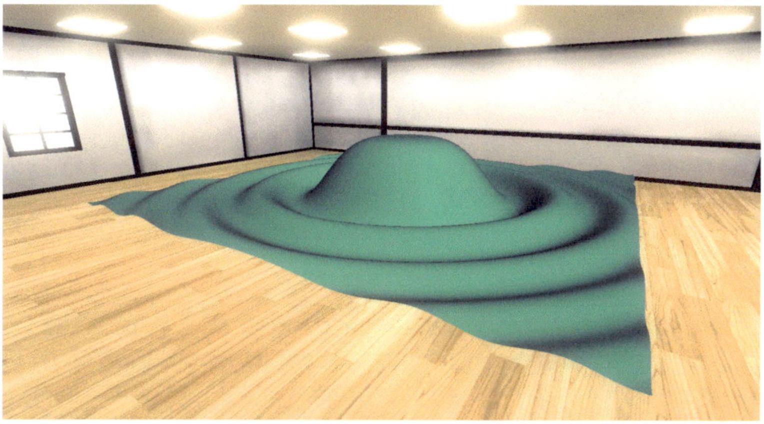

Figure 4.
Dumped sine plot.

2. Related works

Stimulating multiple sensory dimensions during the learning process of a concept or idea makes the learning itself more effective, because our mind needs to experience the reality all around by multiple experiential levels. Indeed, it is essential for us to look, listen and touch in order to build a complex conceptual structure that bears and strengthens our knowledge about reality itself [22, 23]. This multisensorial learning is also important in special-need situations, such as autism [24, 25] or dyslexia [26, 27].

Compared to traditional two-dimensional learning systems, such as a book or the simple screen of a PC or tablet, an immersive study environment has several advantages, as it gives the spatial understanding of the concepts exposed, more peripheral awareness or more useful information bandwidth and decreasing the information dispersion phenomenon [28].

The natural sciences have pioneered the use of Virtual Reality techniques to facilitate the learning of their subjects: chemistry [29–33], biology [34], physiology [35], physics [36].

An effective approach to enhance math skills, knowledge and competences inevitably involves the creation of dynamic digital environments rich in captivating content, which would even reduce anxiety and improve results [37].

Mathematics is known to be one of the most difficult subjects for many students. A recent study asks questions about traditional way of learning and recommends more active and attractive learning approaches [38].

Furthermore, many several studies have shown that immersion in a digital environment can significantly improve education in several ways, because, as mentioned earlier, multiple perspectives are activated at the same time [39]; this fact has an important impact in all areas of mathematics [40].

In some works it has been pointed out that many students find very difficult to relate a function's analytical form to its relative graph, despite the mathematical simplicity of the concept of the function itself [41]; it almost seems that the intermediate layers existing between the set concept and the analytic-geometric one prevent a clear understanding of the link [42, 43].

A recent study has shown that a deeper understanding of the analytical-geometric links existing in functions, if supported by specific software, can help students to develop a positive mindset towards mathematics itself, in terms of attitude, motivation, interest and competence [44]. But we assume for what is the current state of the art that more research is needed to grasp the profound implications that exist.

3. The virtual world made on Unity3D, blender and Vuforia

In this section, we explain how to represent a three-dimensional function within a virtual world created with Unity3D.

The creation of the figures is carried out through two scripts working on different levels:

1. at the beginning it is important to define the resolution, that is the level of detail the figure must have. If for instance we set a resolution equal to 500, then we will have a matrix of 500x500 function values obtained varying X and Y along the grid. The implemented code produces in output the list of vertices that make up the figure.

2. then, the list of triangles must be calculated, since they are the basic elements in computer graphics. Indeed, they specify how the points are interconnected and how they should be represented on the screen.

3. the three-dimensional mesh is generated from the triangle vertices. To do this, a generic script has to be written, which can be called from all the script programs of the Unity3D project.

Once the first phase has been completed, the surface lighting needs to be adjusted, to optimize the appearance of the represented shape, with respect to the user's camera position. This calculation, which seems to be complex, is carried out very quickly in Unity3D. At the end of this operation the mesh is ready to be shown on screen.

4. The architecture of the system

The software used to create the scenarios is Unity3D that allows the composition of virtual environments starting from basic elements which the scene is composed with. It also takes care of rendering, real-time lighting calculation and user interaction management. The fundamental tools that have been used are the following:

- the game objects
- the scripts
- the colliders

The game objects are the basic elements that make up the scene that must be shown to the user.

The scripts are code files written in C# language that permit some predefined tasks to be executed, like managing the appearance of objects on the displayed area, or the camera movement as soon as a key on the keyboard gets pressed.

The colliders are objects that prevent intersection or collision between the character controlled by the user and other different objects in the scene.

Two consecutive C# scripts are tasked with generating all the necessary shapes and modeling any mathematical function in two or three dimensions:

- the first script generates the vertices
- the second script receives in input a list of vertices and draws a three-dimensional figure

Generally, we are interested in drawing mathematical functions with a continuous domain and relative range in $\mathbb{R}$ (Real Number Set); however, this is only an ideal case, since plotting a function through a calculating machine requires simplifying the set of calculations by passing through a discrete space of points. In other words, we need to trace a grid of points that are going to end up defining the level of maximum detail we want to achieve. Also, we must keep in mind that the higher the level of detail, the more calculations the user's device ought to perform to display the object on the screen. A balance must therefore be found between the graphic quality to be obtained and the computational complexity necessary to achieve similar results. In fact, defining a grid of points is equivalent to defining a sampling frequency: if the number of samples is too low, we can in fact obtain" aliasing", getting this way an imprecise representation of the mathematical function we want to show.

In order to improve performance, it was decided to proceed according to this path:

- the shapes generated with Unity3D are saved inside the filesystem so that they can be reloaded directly when the program starts, without the need to recalculate all the objects from the beginning every time

- the shapes have been elaborated with Blender, a software designed for the elaboration of three-dimensional models and objects, in order to reduce the polygonal complexity without modifying the information content

- the elaborated shapes are included again in the Unity3D project

In other words, the complexity of the figures in terms of vertices has been reduced but this is absolutely not perceived by the user who observes them without noticing any difference.

The Augmented Reality (AR) environment uses the Vuforia framework. The program created is an Android Package Kit (APK), which can be installed on Android smartphones with 7.0+ operating system. Vuforia is a Software Development Kit (SDK) that enable the users to analyze the video stream recorded in real time by the phone camera. Vuforia enables the creation of a database of markers (called Vumark), manually associated to the game objects of the scene.

5. Usability of the system

In order to increase the usability and effectiveness of the platform [45], two different modes have been created for user experience: one uses VR only, the other uses AR. In both cases the graphic engine used is the same: Unity3D.

In the Virtual Reality case, a room has been created, and inside the room the three-dimensional figures have been positioned, as shown in **Figure 5**. This software allows the composition of virtual environments starting from basic elements

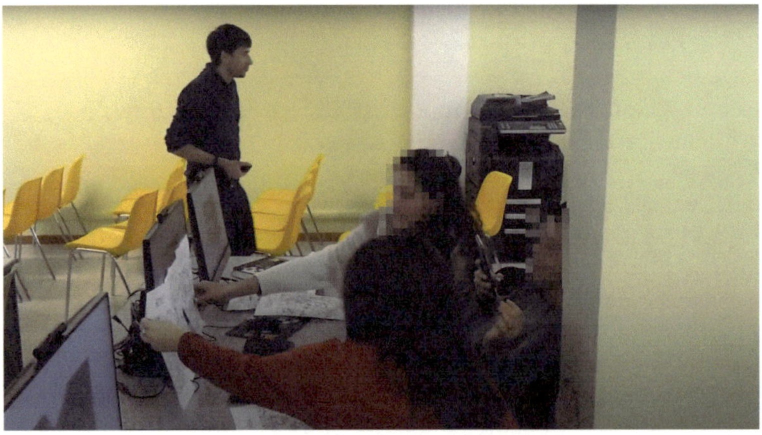

Figure 5.
Students during training session to learn how to use the platform.

called Assets which make up the scene. The VR environment is generated and compiled by WebGL technology: this means that the application is compatible with all devices (computers or smartphones) on the market since the virtual world can be viewed with a web browser, regardless of the operating system used (Windows, Linux, Android, iOS, etc). In particular, in the experience we made for collecting the student's usability evaluations, we focused on smartphones with Android operating system. The graphic quality of the scene adapts according to the computational power of the device, while remaining undemanding in terms of hardware requirements. The scene can be observed through a virtual reality viewer, such as HTC Vive, or through a normal computer monitor. The user has the possibility to move around the virtual environment using the touchscreen on the smartphone, or the mouse and the keyboard on a computer. Inside the environment are visible three-dimensional geometric shapes that support the learning of mathematical functions that are otherwise difficult to be drawn (see **Figures 6** and 7).

As far as the use with Augmented Reality is concerned, Vuforia software has been used. It is a framework integrated in Unity3D and allows to create projects that

Figure 6.
Evaluation of the experience by the students.

Figure 7.
Environment for the VR experience.

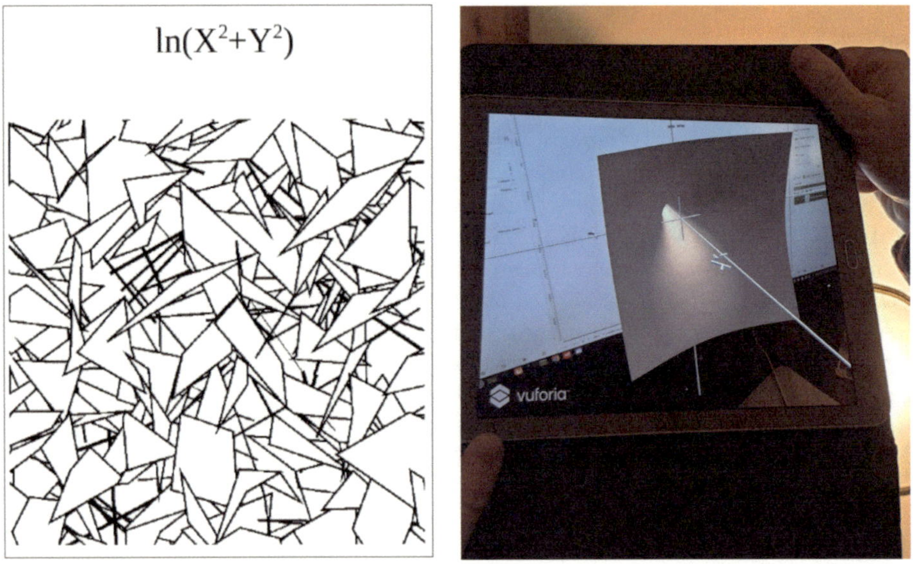

Figure 8.
Example of Vumark to draw the function ln $(x^2 + y^2)$ (on the left) three-dimensional image generated by Vuforia for the object described by the previous Vumarks (on the right).

use Augmented Reality by providing all functions essential for operating on mobile phones. The system is very dynamic, since every time one of the markers in the database is framed by the user's device camera, Vuforia warns Unity to show the object on the scene (and therefore on the user's screen) associated with the framed Vumark. Furthermore, this SDK manages the spatial orientation of the object based on the user's position relative to the Vumark. If we frame a Vumark and move around it, the object associated with it will also rotate, allowing us to appreciate it in a realistic way (see **Figure 8**).

6. Discussion of results

We collected students' feedback to evaluate the usability of the proposed system and to enhance the user's experience, making the AR and VR environments more interactive and attractive. This is the reason why we asked high school students to express their opinion about the quality of the experience, by filling out a questionnaire. Our sample set was composed by 90 high school students, homogeneous by age, gender, social and cultural levels. In **Figure 9** is shown the distribution of the sample per age, while in **Figure 10** is shown the distribution per gender.

The aim of the experiment conducted during a morning class was to obtain a set of coherent and indicative answers on the following main topic: *how much virtual reality and augmented reality can impact the perception of mathematics*. To do that, each class of the school composing the sample, was taken to the computer room, equipped with Vmarks, and left it *playing* with the application on both computers and smartphones for 50 minutes. Finally, the students filled in the questionnaire.

One of the most remarkable aspects of this experience was to observe so clearly and explicitly how the introduction of a play environment stimulated girls and boys in an incredible way. It seemed that the game scenario by itself made them feel comfortable and willing to experience approaching mathematics in a new way.

The results have been being very promising, as most students found the experience of VR and AR applied to some mathematical functions very useful and

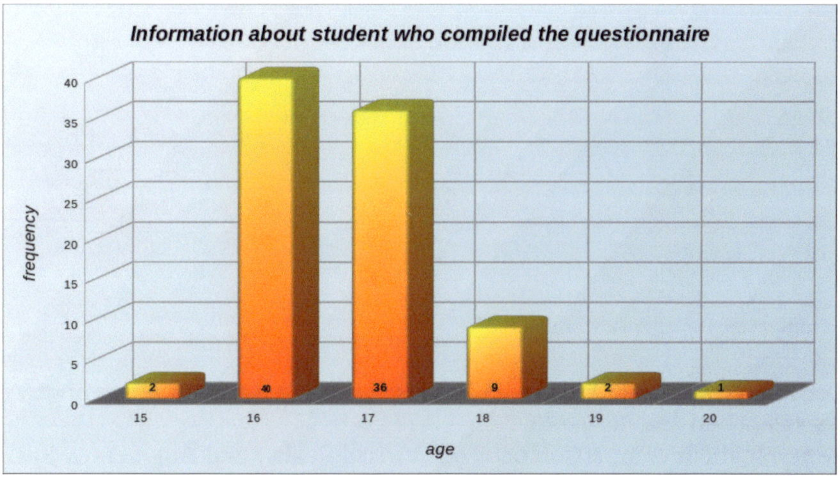

Figure 9.
Distribution by age of the sample of students.

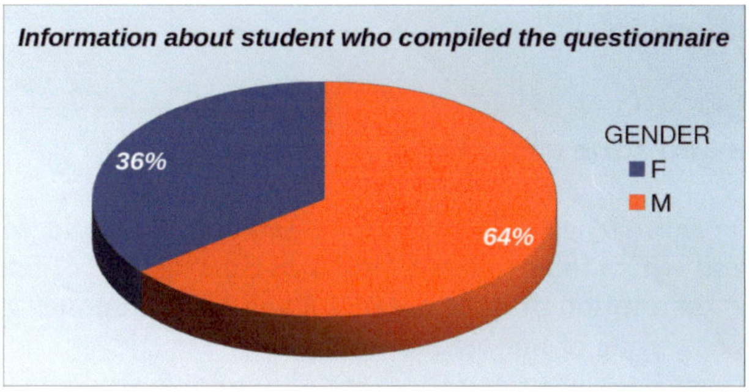

Figure 10.
Distribution by gender of the sample of students.

instructive. Some even asked for an enhancement of the experience, with the addition of new entities and animations.

In **Figure 11** the degree of appreciation of Virtual Reality, i.e. the exploration of the virtual world on a PC, is shown. It appears that the *usability* has been really appreciated, a bit worse were the results in terms of *easy of use*, *user experience* and *graphic quality*.

In **Figure 12** the degree of appreciation of Augmented Reality, i.e. the exploration of the virtual world on a smartphone, is shown. Also in this case the *usability*

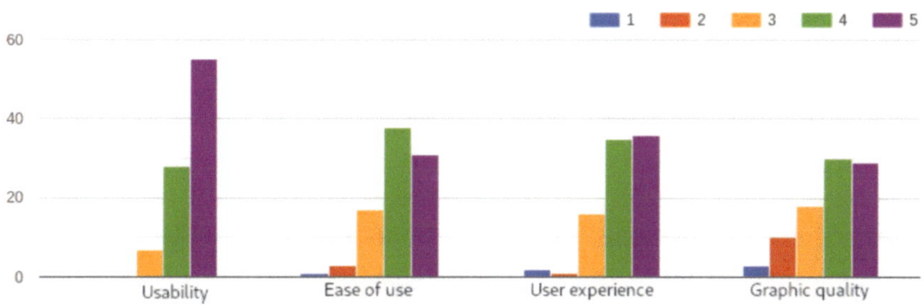

Figure 11.
Degree of appreciation of virtual reality.

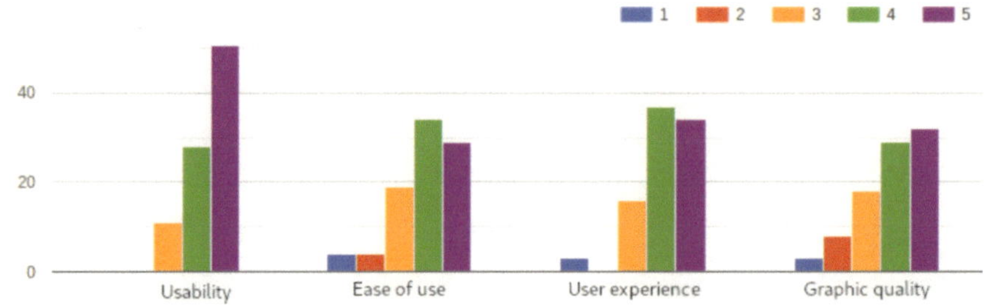

Figure 12.
Degree of appreciation of augmented reality.

has been really appreciated, while a bit worse were the results in terms of *easy of use*, *user experience* and *graphic quality*.

The result about comparison between Virtual Reality and Augmented Reality evaluations are very similar, as it can be seen in the **Figures 11** and **12**.

We think that, if this approach would be largely adopted by teachers, the interest of students on the discipline may significantly raise up and the score of the students may significantly increase, especially in the case of students with problems approaching mathematics.

7. Conclusions and future works

When we first approached our research we had to understand which mathematical objects and which tools to use to enhance the teaching of mathematics, starting from the assumption that the use of AR and VR techniques greatly increases the level of understanding of mathematics.

Our goal has ever been just from the beginning to give students an environment of objects to study and focus on, by selecting among them those of greatest interest and didactic utility for themselves and the entire class. Moreover, we have led our research towards the field of immersive learning, in particular those applications that allow the user to be immersed in virtual worlds in order to increase brain stimulation during the learning phase.

At the same time it has been essential to understand the degree of absorption and emphatic response of students to the system: sensations, disturbances, emotions. At the current state of work, the system can only display objects that are statically compiled: we are working to make the system more dynamic and responsive, which allows the math teacher to draw graphs, two-dimensional or three-dimensional, without the need to print each time a new Vumark, on precompiled functions.

The goal is to get a dynamic platform that lets us understand how the choice of a function and a complete immersive experience in the mathematical object itself (including its specific characteristics and properties) impacts students' learning.

Acknowledgements

Our most heartfelt thanks go to the teachers of the Institute of Higher Education (ITAS) Giordano Bruno of Perugia, Italy for the efficient and exquisite collaboration provided.

A particular thanks goes to the students who have been involved in a very constructive experience and have responded diligently to the questionnaire.

Author details

Damiano Perri[1,2]†, Marco Simonetti[1,2]†, Sergio Tasso[2]† and Osvaldo Gervasi[2]*†

1 University of Florence, Florence, Italy

2 University of Perugia, Perugia, Italy

*Address all correspondence to: osvaldo.gervasi@unipg.it

† These authors contributed equally.

References

[1] Valentina Franzoni and Osvaldo Gervasi. *Guidelines for Web Usability and Accessibility on the Nintendo Wii*, pages 19–40. Springer Berlin Heidelberg, Berlin, Heidelberg, 2009.

[2] Francesca Santucci, Federico Frenguelli, Alessandro De Angelis, Ilaria Cuccaro, Damiano Perri, and Marco Simonetti. An immersive open source environment using godot. In Osvaldo Gervasi, Beniamino Murgante, Sanjay Misra, Chiara Garau, Ivan Blečić, David Taniar, Bernady O. Apduhan, Ana Maria A. C. Rocha, Eufemia Tarantino, Carmelo Maria Torre, and Yeliz Karaca, editors, *Computational Science and Its Applications – ICCSA 2020*, pages 784–798, Cham, 2020. Springer International Publishing.

[3] Gürkan Yildirim, Mehmet Elban, and Serkan Yildirim. Analysis of use of virtual reality technologies in history education: A case study. *Asian Journal of Education and Training*, 4:62–69, 01 2018.

[4] Marco Simonetti, Damiano Perri, Natale Amato, and Osvaldo Gervasi. Teaching math with the help of virtual reality. In Osvaldo Gervasi, Beniamino Murgante, Sanjay Misra, Chiara Garau, Ivan Blečić, David Taniar, Bernady O. Apduhan, Ana Maria A. C. Rocha, Eufemia Tarantino, Carmelo Maria Torre, and Yeliz Karaca, editors, *Computational Science and Its Applications – ICCSA 2020*, pages 799–809, Cham, 2020. Springer International Publishing.

[5] Paul Williams and JS Perry Hobson. Virtual reality and tourism: fact or fantasy? *Tourism Management*, 16(6): 423–427, 1995.

[6] Ryan Yung and Catheryn Khoo-Lattimore. New realities: a systematic literature review on virtual reality and augmented reality in tourism research. *Current Issues in Tourism*, 22(17):2056–2081, 2019.

[7] F. Doil, W. Schreiber, T. Alt, and C. Patron. Augmented reality for manufacturing planning. In *Proceedings of the Workshop on Virtual Environments 2003*, EGVE '03, page 71–76, New York, NY, USA, 2003. Association for Computing Machinery.

[8] T.S. Mujber, T. Szecsi, and M.S.J. Hashmi. Virtual reality applications in manufacturing process simulation. *Journal of Materials Processing Technology*, 155–156:1834–1838, 2004. Proceedings of the International Conference on Advances in Materials and Processing Technologies: Part 2.

[9] A. A. Lazar, W. Choe, K. Fairchild, and Ng Hern. Exploiting virtual reality for network management. In *[Proceedings] Singapore ICCS/ISITA '92*, pages 979–983 vol.3, 1992.

[10] Russell M. Taylor, Thomas C. Hudson, Adam Seeger, Hans Weber, Jeffrey Juliano, and Aron T. Helser. Vrpn: A device-independent, network-transparent vr peripheral system. In *Proceedings of the ACM Symposium on Virtual Reality Software and Technology*, VRST '01, page 55–61, New York, NY, USA, 2001. Association for Computing Machinery.

[11] O. Gervasi, D. Russo, and F. Vella. The aes implantation based on opencl for multi/many core architecture. In *2010 International Conference on Computational Science and Its Applications*, pages 129–134, 2010.

[12] Flavio Vella, Igor Neri, Osvaldo Gervasi, and Sergio Tasso. A simulation framework for scheduling performance evaluation on cpu-gpu heterogeneous system. In Beniamino Murgante, Osvaldo Gervasi, Sanjay Misra, Nadia Nedjah, Ana Maria A. C. Rocha, David

Taniar, and Bernady O. Apduhan, editors, *Computational Science and Its Applications – ICCSA 2012*, pages 457–469, Berlin, Heidelberg, 2012. Springer Berlin Heidelberg.

[13] Panagiota Papadopoulou. Applying virtual reality for trust-building e-commerce environments. *Virtual Reality*, 11:107–127, 06 2007.

[14] Osvaldo Gervasi, Riccardo Magni, and Mauro Zampolini. Nu!rehavr: Virtual reality in neuro tele-rehabilitation of patients with traumatic brain injury and stroke. *Virtual Reality*, 14:131–141, 06 2010.

[15] Mauro Zampolini, Riccardo Magni, and Osvaldo Gervasi. An x3d approach to neuro-rehabilitation. In Osvaldo Gervasi, Beniamino Murgante, Antonio Laganà, David Taniar, Youngsong Mun, and Marina L. Gavrilova, editors, *Computational Science and Its Applications – ICCSA 2008*, pages 78–90, Berlin, Heidelberg, 2008. Springer Berlin Heidelberg.

[16] Osvaldo Gervasi, Riccardo Magni, and Matteo Riganelli. Mixed reality for improving tele-rehabilitation practices. In Osvaldo Gervasi, Beniamino Murgante, Sanjay Misra, Marina L. Gavrilova, Ana Maria Alves Coutinho Rocha, Carmelo Torre, David Taniar, and Bernady O. Apduhan, editors, *Computational Science and Its Applications – ICCSA 2015*, pages 569–580, Cham, 2015. Springer International Publishing.

[17] Osvaldo Gervasi, Riccardo Magni, and Stefano Macellari. A brain computer interface for enhancing the communication of people with severe impairment. In Beniamino Murgante, Sanjay Misra, Ana Maria A. C. Rocha, Carmelo Torre, Jorge Gustavo Rocha, Maria Irene Falcão, David Taniar, Bernady O. Apduhan, and Osvaldo Gervasi, editors, *Computational Science and Its Applications – ICCSA 2014*, pages 709–721, Cham, 2014. Springer International Publishing.

[18] Sheng Bin, Saleha Masood, and Younhyun Jung. Chapter twenty - virtual and augmented reality in medicine. In David Dagan Feng, editor, *Biomedical Information Technology (Second Edition)*, Biomedical Engineering, pages 673–686. Academic Press, second edition edition, 2020.

[19] Damiano Perri, Marco Simonetti, Andrea Lombardi, Noelia Faginas-Lago, and Osvaldo Gervasi. Binary classification of proteins by a machine learning approach. In Osvaldo Gervasi, Beniamino Murgante, Sanjay Misra, Chiara Garau, Ivan Blečić, David Taniar, Bernady O. Apduhan, Ana Maria A. C. Rocha, Eufemia Tarantino, Carmelo Maria Torre, and Yeliz Karaca, editors, *Computational Science and Its Applications – ICCSA 2020*, pages 549–558, Cham, 2020. Springer International Publishing.

[20] Carl B. Boyer Uta C. Merzbach. *A History of Mathematics*. John Wiley & Sons, 3rd ed. edition, 2011.

[21] R. Creighton Buck. *Advanced Calculus*. Long Grove, IL: Waveland Press., 3rd ed. edition, 2003.

[22] Michael Meehan, Brent Insko, Mary Whitton, and Frederick P Brooks Jr. Physiological measures of presence in stressful virtual environments. *Acm transactions on graphics (tog)*, 21(3):645–652, 2002.

[23] Dheva Raja, Doug Bowman, John Lucas, and Chris North. Exploring the benefits of immersion in abstract information visualization. In *Proc. Immersive Projection Technology Workshop*, pages 61–69, 2004.

[24] Sarah Parsons and Sue Cobb. State-of-the-art of virtual reality technologies for children on the autism spectrum. *European Journal of Special Needs Education*, 26(3):355–366, 2011.

[25] Michelle R Kandalaft, Nyaz Didehbani, Daniel C Krawczyk, Tandra T Allen, and Sandra B Chapman. Virtual reality social cognition training for young adults with high-functioning autism. *Journal of autism and developmental disorders,* 43(1):34–44, 2013.

[26] Katerina Kalyvioti and Tassos A Mikropoulos. Virtual environments and dyslexia: A literature review. *Procedia Computer Science,* 27:138–147, 2014.

[27] Elisa Pedroli, Patrizia Padula, Andrea Guala, Maria Teresa Meardi, Giuseppe Riva, and Giovanni Albani. A psychometric tool for a virtual reality rehabilitation approach for dyslexia. *Computational and mathematical methods in medicine*, 2017, 2017.

[28] Doug A Bowman and Ryan P McMahan. Virtual reality: how much immersion is enough? *Computer,* 40(7): 36–43, 2007.

[29] O. Gervasi, A. Riganelli, L. Pacifici, and A. Laganà. Vmslab-g: a virtual laboratory prototype for molecular science on the grid. *Future Generation Computer Systems*, 20(5):717–726, 2004. Computational Chemistry and Molecular Dynamics.

[30] Osvaldo Gervasi, Antonio Riganelli, and Antonio Laganà. Virtual reality applied to molecular sciences. In Antonio Laganá, Marina L. Gavrilova, Vipin Kumar, Youngsong Mun, C. J. Kenneth Tan, and Osvaldo Gervasi, editors, *Computational Science and Its Applications – ICCSA 2004,* pages 827–836, Berlin, Heidelberg, 2004. Springer Berlin Heidelberg.

[31] Antonio Riganelli, Osvaldo Gervasi, Antonio Laganà, and Johannes Froehlich. Virtual chemical laboratories and their management on the web. In Osvaldo Gervasi, Marina L. Gavrilova, Vipin Kumar, Antonio Laganà, Heow Pueh Lee, Youngsong Mun, David Taniar, and Chih Jeng Kenneth Tan, editors, *Computational Science and Its Applications – ICCSA 2005,* pages 905–912, Berlin, Heidelberg, 2005. Springer Berlin Heidelberg.

[32] Antonio Riganelli, Osvaldo Gervasi, Antonio Laganà, and Margarita Albertì. A multiscale virtual reality approach to chemical experiments. In Peter M. A. Sloot, David Abramson, Alexander V. Bogdanov, Yuriy E. Gorbachev, Jack J. Dongarra, and Albert Y. Zomaya, editors, *Computational Science — ICCS 2003,* pages 324–330, Berlin, Heidelberg, 2003. Springer Berlin Heidelberg.

[33] J Georgiou, K Dimitropoulos, and A Manitsaris. A virtual reality laboratory for distance education in chemistry. *International Journal of Social Sciences,* 2 (1):34–41, 2007.

[34] Sandra Tan and Russell Waugh. Use of virtual-reality in teaching and learning molecular biology. In *3D immersive and interactive learning,* pages 17–43. Springer, 2013.

[35] John Ryan, Carol O'Sullivan, C Bell, and R Mooney. A virtual reality electrocardiography teaching tool. In *Proceedings of the Second International Conference, Biomedical Engineering, Innsbruck,* pages 250–253, 2004.

[36] Craig Savage, D. McGrath, Timothy McIntyre, M.J. Wegener, and M. Williamson. Teaching physics using virtual reality. In *AIP Conference Proceedings*, volume 1263, pages 126–129, 10 2009.

[37] Hyesang Chang and Sian L Beilock. The math anxiety-math performance link and its relation to individual and environmental factors: a review of current behavioral and psychophysiological research. *Current Opinion in Behavioral Sciences,* 10:33–38, 2016. Neuroscience of education.

[38] Scott Freeman, Sarah L. Eddy, Miles McDonough, Michelle K. Smith,

Nnadozie Okoroafor, Hannah Jordt, and Mary Pat Wenderoth. Active learning increases student performance in science, engineering, and mathematics. *Proceedings of the National Academy of Sciences,* 111(23):8410–8415, 2014.

[39] Chris Dede. Immersive interfaces for engagement and learning. *Science*, 323(5910):66–69, 2009.

[40] Adriano Pasqualotti and Carla Maria dal Sasso Freitas. Mat3d: a virtual reality modeling language environment for the teaching and learning of mathematics. *CyberPsychology & Behavior*, 5(5):409–422, 2002.

[41] Hatice Akkoc and David Tall. The simplicity, complexity and complication of the function concept. In *PME conference*, volume 2, pages 2–025, 2002.

[42] Daniel Breidenbach, Ed Dubinsky, Julie Hawks, and Devilyna Nichols. Development of the process conception of function. *Educational studies in mathematics,* 23(3):247–285, 1992.

[43] Hatice Akkoç and David Tall. The function concept: Comprehension and complication. *Proceedings of the British Society for Research into Learning Mathematic,* 23(1):1–6, 2003.

[44] Alessandra King. Using desmos to draw in mathematics. *Australian Mathematics Teacher,* 73(2):33–37, 2017.

[45] Benigni Gladys and Gervasi Osvaldo. Human-machine interaction and agility in the process of developing usable software: A client-user oriented synergy. In Maurtua Inaki, editor, *Human Machine Interaction*, chapter 2. IntechOpen, Rijeka, 2012.

Chapter 2

Usability of Computerised Gaming Simulation for Experiential Learning

Nicolas Becu

Abstract

This chapter examines the impacts of computerization of gaming simulations on their usability. Simulation and gaming is an interdisciplinary domain which rallies, among others, the disciplines of education and modelling, and which aim at helping groups of participants to acquire knowledge and skills on complex topics. Gaming simulations can take the form of haptic games or computerised simulations. Yet, the later form may slow down the learning potential for the users. The chapter describes the different types of computerization of gaming simulations. It then examines the effects of computerization, both from the users' perspective (accessibility, captive effect, and flexibility of use) and from the developers' perspective (material, human, and time requirements). Some paths to overcome barriers to experiential learning of computerised gaming simulation are finally presented.

Keywords: simulation and gaming, usability, human–computer interaction, experiential learning, accessibility, free-play

1. Introduction

In the field of simulation and gaming, the problem of software usability has been raised for many years [1–4], perhaps even since the practice of gaming simulation took off in the 1960s [5]. Simulation and gaming is an interdisciplinary domain which rallies, among others, the disciplines of education and modelling, and which aim at helping people to acquire knowledge and skills on complex subjects in which social dynamics are intertwined with technical and/or environmental problems [6–8]. Gaming is the process by which learning takes place. Simulation, on the other hand, is the process used to represent the many interactions, including social interactions, that make up the complex subject being addressed. The range of uses varies widely: city planning, risk management, natural resources management, land use planning and business management are just a few examples. Gaming simulations, understood as tools or artefacts, are used both in teaching and for decision support [9]. In order to illustrate what a gaming simulation can look like, we briefly describe an example of application. LittoSIM is a gaming simulation application used with technicians and decision-makers in coastal cities to help them develop new strategies for coastal flooding risk management [10]. The simulation artefact can be used to simulate coastal flooding that occurs during storms, the extent and intensity of which depends on the coastal defences and the land use development strategies decided by the players. Players take part in gaming sessions during which they select which coastal defence measures to use (based on various economic, regulatory and operational constraints) and adapt how the land in their fictitious urban areas is used from one year to the next, in an attempt to manage this major risk. Experiential learning is achieved both through the various strategies that the players test using the simulation system and through discussions and exchanges of views between the players on the decisions they need to make [11]. This example illustrates how simulation and the gaming process intertwine to create experiences for participants that help them to acquire new knowledge, reflect on a particular situation and develop new skills for dealing with that type of situation.

Between the end of the 1950s and the end of the 1960s, the use of simulation and gaming on the one hand and computer simulation on the other developed concurrently as two ways of approaching decision support in complex situations [12]. Computer simulation focuses on processing data, finding optimal solutions and comparing various typical scenarios. Simulation and gaming focuses on the lived experience (and in particular the emotional and sensitive dimensions of the lived experience) and on the use of communication and collective intelligence to solve a problem based on each other's opinions and find compromises between everyone's interests [13, 14]. The question that arises for designers of these tools, who are aiming to use simulation as a way of facilitating experiential learning, is **whether the use of computer simulation within a gaming simulation artefact slows down, or even restricts, the learning potential for the users**. Although computer technology and simulation methods have evolved, this question remains topical for simulation and gaming practitioners, who in practice articulate this dilemma as a choice that needs to be made at the start of the development phase for a new application between developing a computer game or developing an haptic game (haptic in the sense that it does not involve any human-computer interactions).

The first section presents the issue of computerization in the domain of simulation and gaming and the different types of configuration of computerised gaming simulations. The second section explores the effects of computerisation, both from the users' perspective and from the developers' perspective. The last section presents some recommendations and advances in research to go beyond the limits of usability of computerised applications.

2. Using computers in gaming simulation artefacts

This section first explains the dilemma that arises when deciding whether to use or not computer technology in a simulation and gaming application. Secondly, we examine in more detail the different forms of computerisation used in gaming simulations.

2.1 The computerisation dilemma in simulation and gaming

The study carried out by Crookall et al. [15] analyses human-computer interactions in several situations that use simulation and in which computer technology is used to a greater or lesser extent. Their results show that, during a gaming session, the computer too frequently monopolises users' attention in use cases where the simulation system is more computerised. This has a detrimental effect on social interaction within the user group and, according to the authors, on experiential learning. This early finding was subsequently corroborated by other work. For example, Paran et al. compare two versions of a game they designed for negotiating the siting of gravel pits: a "paper" (haptic) version and a computerised version [16].

"User-friendly, simple and quick to set up, the paper game puts the emphasis on the psychology of negotiation because it insists on interaction and dialogue between the players, bypassing the cumbersome technical aspects. The simplicity of the materials required for this paper version makes it a malleable simulation game that can be easily adapted to the needs and expectations of its organisers. The computerised platform requires more resources but allows the players to manipulate the tools to help the negotiation process. While dialogue is always required, care must nevertheless be taken to ensure that players do not become overwhelmed by the constant stream of information or the technical aspects." Fedoseev makes the same observation, but he also notes that from the point of view of the game's facilitator[1], a computer-based version is more practical in terms of logistics. A computer is the only gaming equipment required, the tasks involved in completing a round of the game are performed more quickly, and the results and data are provided in digital form, which can be more practical for displaying or analysing them [17].

The study by [18] compares 29 use cases of gaming simulations. The artefacts were used with different types of local stakeholders involved in companion modelling[2] processes, either for prospective planning, co-development or consultation purpose. Of these use cases, 21 were workshops involving a role-playing game[3] and the remaining eight used computer simulations (in which all the decision-making is handled by computerised agents). The comparison between these two forms of simulation is based on the opinions (positive or negative) of the workshop participants and of the designers and experts who observed the workshops. More than 300 argumentative elements were collected, classified and analysed.

The results 'summary (**Figure 1**) shows that role-playing games are particularly useful in creating a space for discussion and interaction between participants. Their ability to generate learning among participants is also an important factor, as is, to a lesser extent, their ability to trigger changes in perception of the system being studied. Role-playing games appears to be a particularly user-friendly tool that can be adapted to different types of participants. It has a fun aspect, creating a detachment that facilitates interaction and reduces tension, which other tools do not offer. However, some participants do not embrace the playful dimension of the proposed system. Moreover, in the vast majority of cases, simulation of a single scenario lasts two to three hours, which limits the potential to repeat the simulation and explore a variety of scenarios.

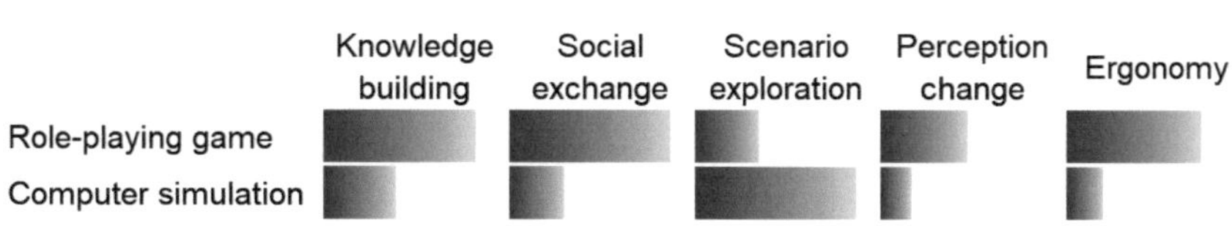

Figure 1.
Comparison of games and computer simulations – Updated version [18].

[1] All the gaming simulation applications discussed here, are implemented during workshops where the players are in attendance and where a facilitator organises and animates the workshop.

[2] Companion modelling is a branch of participatory modelling domain, in which simulation and gaming is widely used.

[3] The term "role-playing game" is used here in the sense used in the literature on companion modelling. A "role-playing game" simulates a real situation involving human participants; it may or may not use computer-based materials but it involves mainly human to human interactions [19]. When computerization is used, it is to represent the decisions of the human players, to record their choices or to display the state of the simulation.

Computer simulation, on the other hand, appears to be particularly well suited to exploring scenarios. A large number of simulations can be run over a short period of time, or even repeated several times, allowing participants to explore different scenarios incrementally [20, 21]. However, it is much less suited to fostering discussion between participants; few changes in perception were noted during the workshops analysed in this study. Their ability to generate learning in the participants is weaker than that of role-playing games, but it still exists. This study thus shows that a computer simulation workshop is rather a space for reflection than a space for social exchange. Lastly, there is a major disadvantage to computer simulation in terms of its poor usability, which can hinder the experience of participants and create a barrier to learning. This is because the computerisation of gaming simulation artefacts tends to reduce their usability and increase their technical sophistication (long waiting times, and difficulty in understanding the content of the tool and in manipulating its interfaces).

2.2 A variety of computerization configurations

Although the previous sections have presented the types of gaming simulations in a somewhat binary way, simply distinguishing between pure computer simulations and non-computer-based games (haptic game, role-playing game,...), in practice there is a whole continuum between computer simulation and haptic games. In the 1960s, Padioleau [22] presented this continuum and classified gaming simulation applications used in the field of political sociology into three categories: those involving only humans (including games for educational use, strategic games for "decision support" and games for theoretical experimentation (e.g. [23])), those involving "mixed simulations" in which (human) participants use computers, and finally "computer simulations". Two decades later, Crookall et al. [15] proposed a classification designed to account for human-computer interactions in a computer simulation environment.[4] This classification distinguishes between two dimensions (**Figure 2**): whether the human or the computer controls the simulation (control of the simulated events and of the overall progress of the simulation) and which type of interaction prevails (human-to-human interaction or human-to-computer interaction).

	Player-to-computer interaction prevails	Player-to-player interaction prevails
Players control the simulation	**Computer-based simulations (CBS)**	**Computer-assisted simulations (CAS)**
Computer controls the simulation	**Computer-dependent simulations (CDS)**	**Computer-controlled simulations (CCS)**

Figure 2.
Classification of human-machine interactions in a computer simulation environment (source: [15], as cited in [17]).

[4] In the Crookall et al. [15] classification, since it focuses on human-computer interactions, games with no computing component are not considered.

The four categories in this classification are as follows:

- Computer-Dependent Simulation (CDS) – A pure computer simulation; participants observe the simulation in the same way as a cinema audience.

- Computer-Controlled Simulation (CCS) – The computer controls the simulation, but the players interact with each other to make decisions when the simulation is interrupted.[5]

- Computer-Based Simulation (CBS) – One or more users interact with the computer continuously as the simulation progresses, for example in a flight simulator.

- Computer-Assisted Simulation (CAS) – Users have roles that are an integral part of the simulation; decisions are made away from the computer and the computer is used solely to perform calculations and record decisions.

This classification provides a meaningful way of understanding the main interaction modes that exist at the interface between simulation and "played simulation". Yet, the ways of interacting with a simulation have evolved since this early classification; technological advances prompt a rethink of the categories proposed by Crookall et al.. In particular, with regard to the CBS category, when it comes to simulations involving several players, todays' technology allows each player to interact individually with a simulation that is shared among several players. In this configuration, human-to-human interactions exist, even though it happens through a computer interface, which usually represents the players in the virtual world by a computer avatar.

Le Page et al. [24] analysed in more detail these inter-player interactions that take place through the computer. To do this, they attempted to characterise the decision-making agents in simulation and gaming artefacts and the types of decision-making agents. The authors consider that the decision can be made either by a human or by a computer program, and that in a played simulation, a human (or a group of humans) can adopt a computer avatar that represents them in the virtual world. They identified four possible types of decision-making agents (from left to right in **Figure 3**):

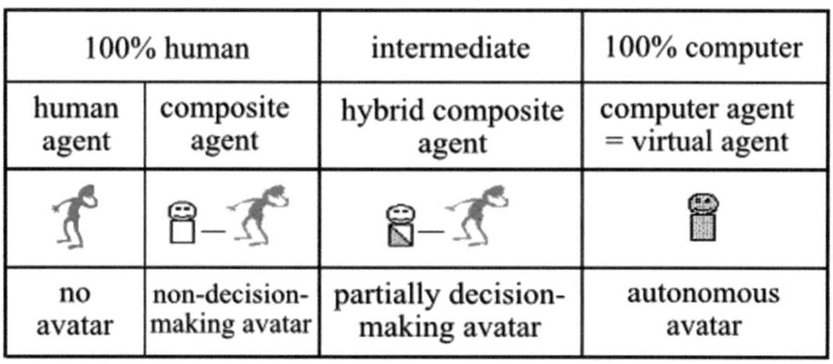

Figure 3.
Types of decision-making agents [24].

[5] Crookall et al. specify that the CCS category relates to simulations in which a group of people interact with each other either to comment on what is happening or to choose a path for the next sequence of the simulation (as in a "choose your own adventure book" but instead of having an individual reader, a whole group of people choose the continuation of the simulation).

(i) the human agent, for whom the decision is 100% human and which has no computer avatar; (ii) the composite agent, for whom the decision is also 100% human but who is represented by a non-decision-making avatar in the virtual world; (ii) the hybrid composite agent, for whom some of the decisions are made by a human and some by a computer program (the computer avatar is then partially decision-making); and (iv) the computer agent whose decisions are 100% derived from a computer program.

To draw a parallel with the previous classification, 100% human composite agent category corresponds to CAS. Conversely, 100% computer agents correspond to CDS, or possibly CCS where the means of control involves something other than the agents. The Le Page et al. classification highlights the range of intermediate configurations that exist between these two end points of the continuum. Within the CBS and CAS categories, there are systems today that include some computer agents and some human or composite agents. In the CBS category, there are also systems that involve only hybrid composite agents. These considerations also relate with ongoing research on hybrid applications, which aims to combine the "space for discussion and social interaction" dimension of games and the "exploratory capabilities" dimension of computer simulations. In this research sector, hybrid game boards for example seek to develop haptic games that use automatic recognition system for in-game actions. Game boards of this kind can be used to design interaction systems between human and computer agents that are much more fluid, or to design new forms of composite agents.

This short literature review shows that the opportunities for interaction between humans through computer technology and digital interfaces have increased significantly, and this raises the question of the link between computer technology and the learning potential of the tool.

3. Effects of computerization in gaming simulations

The first part of this section examines how the degree of computerisation of a gaming simulation affects the user experience, which can lead to inhibiting or promoting experiential learning. Different factors of software usability will be discussed. The second part examines the impact that the use of computers has on the development and deployment of the system. This involves examining the impact that using computer technology has on system designers and simulation workshops' facilitators.

3.1 Weaknesses of computer interfaces' usability for simulation and gaming

The low usability of computerised gaming simulations, as compared to haptic games, has several overlapping causes. These include, the accessibility of the computing environment, captive effect of the computer interfaces and, the flexibility of use of the gaming device.

3.1.1 Accessibility of the computing environment

The computing environment as a medium (screen equipped with a pointing device) is not viewed in the same way by everyone. Some people are more comfortable with computer interfaces than others. For certain groups, this can represent a barrier to the gaming experience. It is important to note that simulation and gaming is practiced with people with a wide variety of backgrounds. For instance, in [20], the participants where Hmong people from northern Thailand and they had no experience whatsoever of using computer screens. In another application, in Grand Morin river basin in France [25], the participants to a second phase of the project were elderly riverside residents who were not familiar with computers. Similarly, in the LittoSIM game, which involves teams of several people, each with a tablet computer, we observed that often the person most comfortable with computers will take care of the tasks that are performed on the tablet. In this case, players are not penalised in relation to the other teams; however, the distribution of roles within the team is open to question. Gourmelon, who tested different types of simulation artefacts (from board games and the more traditional 2D computer simulation to the latest 3D simulation) with different types of participants (scientists, managers and technicians, locally elected officials and school children), notes varying levels of acceptance [26]. Both the school children and locally elected officials fully embrace the 3D simulation, whereas the managers consider it simply as a gadget. She also notes that the level of acceptance of 2D computer simulations varies. Scientists and managers accept the 2D game more easily because they are used to working with these tools in their professional life. By contrast, the school children and locally elected officials find the 2D computer simulation too technical and insufficiently engaging compared to a haptic game.

3.1.2 Captive effect

In the literature of serious game studies, various factors have been identified that help to engage participants and prolong the learning experience: graphic aesthetics and the soundscape; the fluidity of the user experience, allowing the player to lose themselves in the game and forget about the outside world; the narration, which helps to maintain suspense and makes the player want to continue playing; the right level of difficulty and challenge, which maintains the player's concentration and motivation; and the captive effect of the interfaces. This captive effect that a computer interface (whether in a computer game or a digital interface in general) has on its user can be stronger or weaker depending on the person [27]. Researchers in the field attribute this captive effect to two aspects of the computer interface. Firstly, the screen itself *"contributes to drawing our attention towards the screen. [... It] paradoxically forms a 'boundary frame' that restricts our visual perception [and] in a certain way immobilises our gaze, creating a centring that explains why we feel as if we are absorbed, even hypnotised by the screen"* [28]. Secondly, the computer interface has the unique characteristic of juxtaposing different types of visual information – *"the screen is a frame (interface) that contains other frames"* [28] – and this tends to monopolise our attention: *"It then functions as a capture device: we become absorbed, captivated by light, writings, images [...]"* [28]. In fact, in a computerised game, where game design calls for both human-machine interactions and social interactions, the computer interface tends to take up too much of the participants' attention to the detriment of direct exchanges between people.

3.1.3 Flexibility of use and free-play

An important aspect of simulation and gaming is the game-play flexibility. The function of flexibility is to make it easier for the player to take a playful attitude [29], in the sense that the gaming system will be able to conform to his choices and freedom of decision, and not force them. Game-play flexibility aim is to guarantee freedom of action, which must not be compromised by problems related to understanding the interfaces or by technical difficulties. Klabbers argues that the free-play dimension is part of the very specificity and morphology of simulation and gaming devices [8]. Free-play is the idea that users are free to play as they wish.

The path of the simulation is never fully scripted in advance and it is impossible to say what the outcome of the simulation will be before it is played. In terms of game-play flexibility, haptic games have a clear advantage over computer interfaces because of their very nature as tangible objects which can be touched, grasped and manipulated in any desired way. They can be handled with a degree of spontaneity that Duke identifies as an essential element in his definition of gaming/simulation as a mode of communication capable of understanding the "gestalt"[6] [7]. In haptic games, there is a degree of flexibility in using tokens and other tangible objects that is not found in computer interfaces [30]. The user can pick up the token, touch it, examine it in its entirety, whereas in the computer interface, there is always something hidden, symbolically speaking, that the user cannot touch. To put it another way, in a computer game handling is made through an interface built by someone else, whose logic and meaning may not readily accessible to the user, or may even impose itself to the user [2]. In addition, the physical pieces of a haptic game can be used more easily as a medium for communication between players. The players may designate a token or a space on the game board to inform other players of a particular situation or signal their intention. It is possible to provide this type of signalling and communication mechanism with a computer interface, as long as the interface is visible to all, for example by using a horizontal projection surface. Similarly, there is immediacy of action with a token, which is not always the case with computers, especially when several calculations are performed after a player's action. Lastly, the computer interface is developed according to the game mechanics devised by the game designers. If players want to perform an action that has not been coded in advance, they will have difficulty doing it by themselves because that action was not intended. They will first consult the game facilitator or, as the action cannot be performed immediately, will give up attempting that action. Physical game-playing components do not present any such obstacle for players. Players can pick up a game piece and use it for some other purpose or make it do something that was not intended. They can create new game mechanics spontaneously, such as hiding counterfeit money, substituting or adding game pieces, hindering the access of certain players to game resources by physical obstacles – these are all possible ways of hijacking game mechanics that are difficult to reproduce in computerised games.

3.2 Impact on application development and deployment

When the development of a new gaming simulation application begins, a recurring question arises, which the choice between a non-computerised and a computerised game (CAS, CBS or other types of hybrid configurations). To make this choice, the developers will consider the usability factors mentioned earlier, discuss them according to the target audience, and these are weighed up against the required computing capacity. But these are not the only aspects that need to be considered when choosing one type of gaming device over another. Using computer technology also has an impact on development needs, and on the organisation of the gaming workshops. This section examines how easy and difficult it is to develop and deploy (set up and or organised gaming sessions) in relation to their degree of computerisation. The impact of the use of computer technology is examined from three angles: the material and equipment requirements for organising a workshop, the human requirements during the development stage and during the

[6] Duke defines "gestalt" as a structure or configuration of physical, biological or psychological phenomena so intertwined that it constitutes a functional unit whose properties are not deductible from the sum of its parts [7]. The concept of "gestalt" shares properties with the modern concept of a complex system, with the difference that it fully integrates the "human factor" in its definition of the functional unit.

implementation of a workshop, and the impact on development time and workshop time. Before discussing the case of a computerised system, the following paragraph briefly presents the case of a haptic game from these three angles.

Non-computerised games require little in the way of technical equipment. They use game boards, game pieces, cards or other game-playing components. Although the equipment is not technical in nature, some games may involve a large number of components. Non-computerised games generally require significant human resources during their implementation (facilitators, assistants, observers, etc.). Very few games can be played by just one person; where they can, the facilitator is under considerable pressure. Preparing for a game session can either be quick (10 minutes) or require a much longer set-up time (1 hour or even 1.5 hours), depending on the game-playing components required (boards, game pieces, cards, etc.) and how the play area needs to be configured (arrangement of tables and chairs and separation of areas). A game, excluding debriefing, can last from approximately 40 minutes for the fastest games to several hours for slower games (2 hours on average).[7]

3.2.1 Impact on the material requirements for organising workshops

The computerisation of gaming systems has a significant impact on the technical and computer equipment required to organise a workshop. Some systems require equipment that cannot be transported; in these cases, face-to-face game sessions are held in a dedicated room. This is the case, for example, for games that use specially designed interactive tables. This transport constraint does not apply to online computerised gaming simulation which we do not discuss further in this chapter.

When the game is played face-to-face, the computer tools and equipment are such that all the required gaming kit can now be transported in a wheeled suitcase. For example, the LittoSIM game kit comprises several computer terminals, a video projector and a computer server, all of which can be used to organise game sessions involving several teams, each with a dedicated computer terminal and different projection areas [10]. The downside is the installation time (positioning and connecting the equipment and starting the software applications), which takes longer the more computer hardware there is. However, although the installation time can be significant, setting up the simulation is in principle fairly straightforward compared with a non-computerised game. Setting up the non-computerised simulation involves positioning the board and arranging all the game-playing components before the game can begin. For example, the Maritime Spatial Planning board game [31], setting up all the materials takes a good hour.

3.2.2 Impact on human requirements

In terms of human resources, the development of computerising gaming systems requires computer skills in addition to the game design skills required for any type of gaming system. Modelling platforms adapted to gaming simulation development offer dedicated interfaces to simplify the development process. Nevertheless, game designers must know how to use the platform. The benchmark platform for agent-based modelling dedicated to gaming simulation is Cormas [32, 33].

In addition to the needs at the development stage, computer-assisted games may also need a computer operator to be present during the gaming session itself. If the

[7] A workshop's duration (which includes briefing, game/simulation and debriefing) can also vary greatly. Some can be quite short (about 1.5 hours), others last half a day, a day or even several days, particularly when the workshop includes several games and several debriefings.

interface usability and the game design allow, entering data into the computer during the game can be carried out by a facilitator dedicated to this task. For example, in the Djolibois game, a "registration office" is situated at each end of the play area, with one computer and one operator (a person from the organising team) at each "registration office". The player informs the operator of their decisions, the quantities of wood cut at the "forest office" and the quantities of wood sold at the "town office". The operator enters the data into the computer model and informs the player of the results before moving on to the next player [34].

3.2.3 Impact on development time and game time

Computerised systems usually require more development time than their non-computerised counterparts. Yet, the amount of time required varies considerably and the development of computerised gaming simulation can range from just a few weeks to more than a year.

Regarding the game time, computerisation reduces drastically the simulation time, particularly for performing calculations and data update. In many cases, the processing carried out by the computer during the calculation phases would be impossible without computerisation. Although the calculations and updates are faster, this does not necessarily mean that the games played with computerised systems are any shorter. There are computerised gaming systems where games last just as long as those of non-computerised systems (2 to 3 hours on average). There are also gaming systems whose games are very fast. This is particularly the case in configurations where several rounds of the same game are played during the same workshop. Each round simulates a scenario: the first round allows players to familiarise themselves with the game, while subsequent rounds may be played faster (sometimes they may last no more than 20 minutes) and allow to test several contrasted scenarios [35].

4. Paths to overcome barriers to experiential learning

The previous sections have provided a better understanding of how the computer environment affects the gaming experience and can be a barrier to experiential learning. In this section, we present points of attention and some ideas to overcome these barriers.

First, it is important to take into account the type of public targeted and the level of accessibility required for this public. Depending on the target audience, the development of a gaming simulation can aim either for a computerised or a haptic version. The other important factor to take into account is the degree of free-play that is expected in the game. This depends on the objective of use of the gaming simulation. If a high degree of flexibility is to be achieved, it is advisable to develop a haptic version. However, the accessibility and flexibility of a computerised application can be improved, in particular by playing on its ergonomics. This will be further developed in the following of this section. The last part of this section will present some of the advantages of computerised devices, which can be reached when the constraints linked to computerization have been lifted.

4.1 Avoiding excessive technological sophistication and recent advances

The examination of the issues surrounding the usability of computer interfaces has served to highlight the obstacles it can pose to the forms of interaction and communication that are an integral part of the gaming experience [7, 29]. For the configuration of computerised games (computer-assisted games, computer-based games or other types of hybrid configurations), it is therefore essential to avoid excessive technical sophistication and to focus on the usability of the interfaces, paying particular attention to processing time, the clarity of the interfaces and their controllability [1, 10].

The developers of the Cormas simulation computer program, the benchmark multi-agent platform for participatory modelling and simulation, have paid close attention to this question of usability and flexibility of use of the game-playing components and the controllability of the game mechanics [36]. In the past years, they have integrated relatively user-friendly tools for moving and manipulating virtual game pieces into the platform [32]. With several other practitioners, they are now interested in designing hybrid boardgames that would allow players to physically manipulate the game pieces, but calculating the effect of their actions would be computer-based. Such a hybrid boardgame would be a great step forward in overcoming the problems of accessibility and captive effect described above.

Concerning the improvement of the free-play capabilities of computerised gaming simulation, the Cormas developers seek to enhance the control that players can have on the definition of game mechanics. The question, from a modelling point of view, boils down to achieving *"a tighter coupling between the conceptual model and the simulation model by using tools to manipulate both internally"* [19]. Two avenues are explored to this end. Firstly, Bommel integrated tools into the platform that can be used during the game to modify (fairly easily and quickly) the computing specification of the interaction mechanisms [37]. A game facilitator can use these tools fairly easily, but players find it more difficult to use them. Secondly, Christophe Le Page explored the process of gradually creating specifications for the interaction mechanisms with the participants over the course of a simulation [38].

4.2 Taking advantage of computing capabilities

When the constraints linked to their use have been lifted, the features of computerised games can be useful tools, both to encourage participants to reflect on how the system represented functions and to explore potential future scenarios. Compared with non-computerised games, computerised games have four major advantages. First, the use of computer technology means that important and useful calculations can be performed during the game to report on complex physical phenomena or to simulate automatic game actions, for example. Second, this computing capability can also be used at the end of the game to explore different development trajectories, as in the game FisHcope [39]. Third, computer interfaces can be used to represent a large amount of information, especially in different forms, which is particularly useful when it comes to integrate asymmetric information and points of view, distributed among the different players [13]. Lastly, computers reduce the time required to reset the simulation environment between two rounds of the game, because this is done automatically. With a board and game pieces, resetting is done manually and can take several minutes [40]. In some cases, this reduces the number of facilitators needed to run the game. This is the case, for example, with the games Motte-Piquet and Djolibois [34, 41], which require only one facilitator thanks to their user-friendly computer interface for entering players' actions. The initial versions of these games, however, required three or more facilitators.

5. Conclusion

This chapter on the usability of computerised gaming simulations has provided a better understanding of how the computer environment can be a barrier to experiential learning and how these barriers can be overcome. The accessibility and flexibility of use of computer interfaces are two key aspects that need particular attention for usability of computerised applications. The use of computer technology also has an impact on the teams developing and deploying the systems. The choice between designing a haptic system or a computerised system therefore depends on the resources available, the calculation requirements, the display requirements and more specifically the asymmetrical display requirements, and the degree of free-play that needs to be integrated into the system.

The criteria for making this choice will most certainly change as technology develops and as the boundary between these two types of system becomes blurred. The current developments in hybrid boardgames, which mix physical manipulation and digital display, have already been mentioned above. Other innovative forms of human-machine interaction are also beginning to be used in simulation and gaming, including the ability to interact with several people using the same simulation through different individual devices such as tablets or smartphones, or the ability to interact as a group through an interactive table [42]. The development of these new forms of interaction will certainly shake up perceptions of the role that computing plays in gaming simulations.

Author details

Nicolas Becu
National Centre for Scientific Research – CNRS/UMR LIENS 7266, La Rochelle, France

*Address all correspondence to: nicolas.becu@cnrs.fr

References

[1] Jones, K., 1991. Using Computer-Assisted Simulations and Avoiding Computer-Hindered Simulations. Simulation & Gaming 22, 234-238.

[2] Myers, D., 1999. Simulation as play: A semiotic analysis. Simulation and Gaming 30, 147-162.

[3] Thavikulwat, P., 1991a. Modeling the Human Component in Computer-Based Business Simulations. Simulation & Gaming 22, 350-359.

[4] Thavikulwat, P., 1991b. Rebuttal to Wolfe's Comments on Modeling the Human Component in Computer-Based Business Simulations. Simulation & Gaming 22, 365-367.

[5] Klabbers, J.H.G., 2009a. The Saga of ISAGA. Simulation & Gaming 40, 30-47.

[6] Crookall, D., 2012. The founding of modern simulation/gaming: S&G and ISAGA four decades on. Simulation & Gaming 43, 5-14.

[7] Duke, R.D., 1974. Gaming: The future's language. SAGE, London.

[8] Klabbers, J.H.G., 2009b. The magic circle: Principles of gaming & simulation. Sens publishers, Rotterdam, The Netherlands.

[9] Mayer, I.S., 2009. The gaming of policy and the politics of gaming: A review. Simulation and Gaming 40, 825-862.

[10] Becu, N., Amalric, M., Anselme, B., Beck, E., Bertin, X., Delay, E., Long, N., Marilleau, N., Pignon-Mussaud, C., Rousseaux, F., 2017. Participatory simulation to foster social learning on coastal flooding prevention. Environmental modelling and software 98, 1-11.

[11] Becu, N., Amalric, M., Anselme, B., Beck, E., Bergossi, P., Delay, E., Marilleau, N., Pignon-Mussaud, C., Rousseaux, F., 2019. Participatory simulations with decision makers on coastal flooding prevention: what did they learn?, in: ISAGA Conference. Warsaw.

[12] Meadows, D.L., 2000. Learning to be simple: My odyssey with games. Simulation and Gaming 30, 342-351.

[13] Becu, N., Frascaria-Lacoste, N., Latune, J., 2014b. Distributed Asymmetric Simulation - Enhancing Participatory Simulation Using the Concept of Habitus, in: Kriz, W.C., Eiselen, T., Manahl, W. (Eds.), ISAGA 2014: The Shift from Teaching to Learning: Individual, Collective and Organizational Learning through Gaming Simulation. Bertelsmann, Dornbirn, Austria, pp. 75-85.

[14] Kriz, W.C., 2003. Creating Effective Learning Environments and Learning Organizations through Gaming Simulation Design. Simulation and Gaming 34.

[15] Crookall, D., Martin, A., Saunders, D., Coote, A., 1986. Human and computer involvement in simulation. Simulation & Gaming 17, 345-375.

[16] Paran, F., Sébastien, L., Graillot, D., 2010. Simulation pédagogique pour l'aide à la négociation environnementale : d'un jeu de rôle de papier à une plate-forme informatisée *[Pedagogical simulation for environmental negotiation support: from a paper role-playing game to a computerised platform]*, in: La Branche, S., Milot, N. (Eds.), Enseigner Les Sciences Sociales de l'environnement ; Un Manuel Pluridisciplinaire. Presses Universitaires du Septentrion, pp. 127-162.

[17] Fedoseev, A., 2016. Comparing Live-Action and Computer-Mediated Educational Games for Engineering

Studies, in: Simulation and Gaming in the Network Society. Springer, pp. 107-119.

[18] Becu, N., Bommel, P., Botta, A., Le Page, C., Perez, P., 2014a. How do participants view the technologies used in companion modelling?, in: Etienne, M. (Ed.), Companion Modelling: A Participatory Approach to Support Sustainable Development. Springer, Versailles, pp. 1-403.

[19] Le Page, C., 2017. Simulation multi-agent interactive : engager des populations locales dans la modélisation des socio-écosystèmes pour stimuler l'apprentissage social. *[Interactive multi-agent simulation: engaging local people in the modelling of socio-ecosystems to stimulate social learning]*. Université Paris 6.

[20] Becu, N., Neef, A., Schreinemachers, P., Sangkapitux, C., 2008. Participatory computer simulation to support collective decision-making: Potential and limits of stakeholder involvement. Land Use Policy 25, 498-509.

[21] Lidon, B., Lopez, J.-M., Sosiawan, H., Kartiwa, B., Triomphe, B., Jamin, J.-Y., Farolfi, S., Bourgeois, R., Becu, N., 2018. Approach and impact of a participatory process for the reorganization of irrigation management: a case study in Indonesia. Cahiers Agricultures 27.

[22] Padioleau, J.-G., 1969. Note sur les "Simulations en sociologie politique" *[Note on "Simulations in Political Sociology"]*. Revue Française de Sociologie 10, 201-219.

[23] Schelling, T.C., 1961. Experimental Games and Bargaining Theory. World Politics 14, 47-68.

[24] Le Page, C., Abrami, G., Barreteau, O., Becu, N., Bommel, P., Botta, A., Dray, A., Monteil, C., Souchère, V., 2014. Models for sharing representations, in: Etienne, M. (Ed.), Companion Modelling. A Participatory Approach to Support Sustainable Development. Springer, pp. 69-101.

[25] Carré, C., Haghe, J.-P., de Coninck, A., Becu, N., Deroubaix, J.-F., Pivano, C., Flipo, N., Le Pichon, C., Tallec, G., 2014. How to integrate scientific models in order to switch from flood control river management to multifunctional river management? International Journal of River Basin Management 12, 231-249.

[26] Gourmelon, F., 2017. Des formes de simulations multi-agents adaptées à différents contextes sociaux d'apprentissage *[Forms of multi-agent simulations adapted to different social learning contexts]*, in: Rencontres Jeux et Enjeux. Université de La Rochelle, La Rochelle.

[27] Frau-Meigs, D., 2011. Penser la société de l'écran. Dispositifs et usages *[Thinking about the screen society. Devices and uses]*, Communication & langages. Presses Sorbonne Nouvelle, Paris.

[28] Seux, C., 2014. Écran(s) *[Screen(s)]*. Le télémaque 45, 15-25.

[29] Brougère, G., 1999. Some elements relating to children's play and adult simulaton/gaming. Simulation & Gaming 30, 134-146.

[30] Abrami, G., Ferrand, N., Ducrot, R., Morardet, S., Hassenforder, E., Garin, P., Robin, P., Aquae-gaudi, W., 2016. Paper and pebbles simulations and modelling for the governance of socio-environmental systems: a review of 8 years of experimenting with the Wat-A-Game toolkit, in: 8th International Congress on Environmental Modelling and Software.

[31] Keijser, X., Ripken, M., Warmelink, H., Abspoel, L., Fairgrieve, R., Mayer, I.S., 2018. Maritime Spatial Planning – A Board Game for Stakeholder

Involvement. Lecture Notes in Artificial Intelligence 10825.

[32] Bommel, P., Becu, N., Le Page, C., Bousquet, F., Page, C. Le, 2015. Cormas, an Agent-Based simulation platform for coupling human decisions with computerized dynamics, in: Kaneda, T., Kanegae, H., Toyoda, Y., Rizzi, P. (Eds.), ISAGA 2015: Hybrid Simulation and Gaming in the Network Society, Translational Systems Sciences. Springer, Singapore, pp. 1-25.

[33] Le Page, C., Becu, N., Bommel, P., Bousquet, F., 2012. Participatory agent-based simulation for renewable resource management: The role of the cormas simulation platform to nurture a community of practice. Journal of Artificial Societies and Social Simulation 15, 10.

[34] Gazull, L., Gautier, D., Becu, N., 2010. Usage d'un jeu de rôles pour l'analyse préalable d'un SIG: DJOLIBOIS, un jeu spatialisé pour l'approvisionnement en bois-énergie de la ville de Bamako (Mali). *[Use of a role-playing game for the preliminary analysis of a GIS: DJOLIBOIS, a spatialized game for the wood energy supply of the city of Bamako (Mali)]*. Revue internationale de géomatique 20, 7-36.

[35] Becu, N., Amblard, F., Brax, N., Gaudou, B., Marilleau, N., 2015. How to Involve Stakeholders in the Modeling Process, in: Agent-Based Spatial Simulation with Netlogo - Volume 1. Elsevier, pp. 223-252.

[36] Bommel, P., Becu, N., Bonté, B., Delay, E., Le Page, C., Morales, H., Papoulias, N., Stinckwich, S., 2018b. CORMAS, a participatory and interdisciplinary Agent Based Simulation Platform, in: European Smalltalk User Group. Cagliari, Italy, p. 2018.

[37] Pierre Bommel , Francisco Dieguez , Danilo Bartaburu , Emilio Duarte , Esteban Montes , Marcelo Pereira Machín , Jorge Corral, C.J.P. de L. and H.M.G., Bommel, P., Dieguez, F., Bartaburu, D., Duarte, E., Montes, E., Machín, M.P., Corral, J., De Lucena, C.J.P., Grosskopf, H.M., 2014. A Further Step Towards Participatory Modelling: Fostering Stakeholder Involvement in Designing Models by Using Executable UML. Journal of Artificial Societies and Social Simulation 17, 1-9.

[38] Le Page, C., Bobo, K.S., Kamgaing, T.O.W., Ngahane, B.F., Waltert, M., 2015. Interactive Simulations with a Stylized Scale Model to Codesign with Villagers an Agent-Based Model of Bushmeat Hunting in the Periphery of Korup National Park (Cameroon). Journal of Artificial Societies and Social Simulation 18, 3-11.

[39] Berry, A., Becu, N., Lapijover, A., Deysson, H., 2019. Expérimenter par un jeu sérieux des compromis entre acteurs à travers l'usage des savoirs locaux pour une gestion durable des écosystèmes marins *[A serious game to integrate fishermen local knowledge and experiment stakeholders' compromises for sustainable management of marine ecosystems]*, in: OCEANEXT. Nantes, France.

[40] Bommel, P., Becu, N., Bonté, B., Delay, E., Le Page, C., 2018a. Cormas in 10 years!, in: CoMSES Net Virtual Conference.

[41] Gourmelon, F., Rouan, M., Lefevre, J.-F., Rognant, A., 2011. Role-Playing Game and Learning for Young People About Sustainable Development Stakes: An Experiment in Transferring and Adapting Interdisciplinary Scientific Knowledge. Journal of Artificial Societies and Social Simulation 14, 21.

[42] Marty, J.-C., Serna, A., Carron, T., Pernelle, P., Wayntal, D., 2016. Multi-device Territoriality to Support Collaborative Activities: Implementation and Findings from the E-Learning Domain, in: 11th European Conference on Technology Enhanced Learning. pp. 152-164.

Section 2

Breaking Down Barriers

Chapter 3

Smartphones for Vision Rehabilitation: Accessible Features and Apps, Opportunity, Challenges, and Usability Evaluation

Suraj Singh Senjam

Abstract

In today's world, digital technology and smartphones have become a part of our everyday lives. Smartphones are one of the most advance forms of digital technology that can be viewed as assistive technology for disabled, including for individual with a visual loss. However, they are often not considered as assistive technology for visual impairment and blind, especially in low middle-income countries. There have been a lot of development in the mobile technology that incorporates computer technology, including electronic information, communication technology as well as touch-screen accessibility. Such an advancement in smart technology of mobile devices leads to the transformation of the interface technique from visual smartphones interaction into a truly eyes-free means interaction by using other body senses, such as haptic, gesture, and sound, etc. These innovative accessible features and applications enhance the accessibility of smartphones significantly to individuals with visual impairment. There are many built-in accessible features and third-party accessible applications that enable to access many useful information and contents in the smartphones. Such aesthetic technology facilitates in performing daily activities, independent functioning, movement, social inclusion and participation, educational activities, accessing information of today's digital society, sighted help, and finally helps to improve the quality of life. Therefore, these smart technologies make smartphones to serve as assistive technology for people with visual impairment and blindness. The smartphones are visually and physically demanding, and are ubiquitous any time and any place, and user can carry it at everywhere. They are universally design, so less social stigma to the users and less discomfort when using it. To view smartphones as assistive technology universally, healthcare providers, caregivers or rehabilitation professionals need to be informed, and make aware of the beneficial aspect of smartphones and its accessibility. Finally, engineers and developers are continuously fostering to develop more innovative and readily accessible apps for visual impairment. Since single app does not fit all purposes for visual impairment and blind, there is a potential need of developing clinical guidelines on the use of such accessible apps or features that will help to recommend appropriately for various types of functions.

Keywords: smartphones, accessible applications, visual impairment, usability evaluation, vision rehabilitation

1. Introduction

The World Health Organization classifies the blindness as a severe disability that corresponds with its category VI, whereas the category VII is considered to be a maximum severity [1]. Visual impairment and blindness are problems which can affect significantly in functioning for daily living activities, to live with independent life, indoor and outdoor movement, social inclusion & participation, communication, employment, and finally impacted on the quality of life [2–4]. Such impacts extend much beyond individuals who have the problem, but also to the family, to the society as well as the community to a large extent. In addition, the present COVID-19 pandemic and its preventive measures pose a new challenge in terms of performing the daily living tasks among visually disabled people as well as to receive their daily supports for living [5]. Globally, around 253 million people who have some form of visual loss are facing such challenges and difficulties in their everyday life [6]. These people need to live with independent lives, and cope these daily challenges and difficulties resulting from visual impairment whether it is at home, workplaces, schools, or market. Fortunately, the continued advance in assistive technology has provided a new platform and opportunities for people living with visual impairment to overcome many of these barriers and challenges that they encounter in their everyday lives. Digital assistive technology is one of them that has grown rapidly in the past few years which helps in solving these challenges.

Many smart digital assistive technologies based on electronic information, communication technology for visual impairment have been gaining a lot of importance across the world in recent times [7, 8]. Such smart assistive technologies have many accessible features and accessible applications for persons with disabilities, including low vision and blindness. For instance, the technology of mainstream assistive devices, e.g., mobile phones and tablets, has evolved substantially over the years from simple basic phone, such as NOKIA 8110, to high end and touch screen smartphones or tablets or I-pad with operating system of IOS, or Android platform (**Figure 1**). With ongoing advance in smartphone technologies, it is becoming even more feasible for the person with visual impairment to rely on mobile technology in understanding their immediate surrounding, and to access huge amounts information that can improve their level of independent functioning, movement, social inclusion, participation, educational activities, and finally helps in improving the

Figure 1.
Simple phone, smartphone, and tablet (left to right).

Figure 2.
Braille book, Braille slate and stylus, Braille type writer (left to right).

quality of life [9]. Now, smartphones have become a part of our everyday life and replacing gradually the traditional assistive devices (such as Braille materials **Figure 2**) in doing various routine tasks and bringing the solutions [10, 11].

2. Smartphones assistive technology for persons with visual impairment

Conventionally, people and even health care providers, frequently, do not consider smartphones as assistive technology for visual impairment and blindness [12, 13]. Since, the smartphone technology heavily relies on a good visual function, many believe that such a technology is not particularly accessible to individuals with visual impairment. There is perceived limitation on the use of smartphones among the general public, eye care professionals, and caregivers, by visually impaired people.

Over the last few years, the research on "Human-Computer -Interaction" has been gaining a lot of attention, and have been developed many innovative assistive technologies that drive with a new interface design which not only makes more accessible, user-centered but also user-friendly for visually disabled people. Such

Figure 3.
Eyes-free smartphone interaction using gesture.

remarkable changes in assistive technology development transform the need of visual interaction to audio or haptic or tactile interaction with the technology. Till now smartphones are one of the most advance features of computer technology of the 21st century, that contain many functionalities of advance computer technologies, including technologies related to electronic information and communication system. Various innovations on usage of other body senses, for example, gestures, haptic, or audio, other than vision are being developed, making a truly eyes-free means in human smartphone interaction (**Figure 3**), and thereby, significantly improve accessibility to individuals with visual impairment. The smartphones have a large touchscreen system to run it, and can access mobile internet data, Wi-Fi and Bluetooth. Such an innovative touch-screen accessibility leads to an obvious benefit to visually challenged individuals, and also helps to overcome many challenges. The common operating systems of smartphones are based on the Android, Apple iOS and Windows phone, BlackBerry; Android being the most common, 59.1% in market shared [14].

3. Accessible built-in features for visual impairment and blind

There are built-in accessible features in smartphones that facilitate for all users regardless of disabilities or individuals' functional limitations. These accessible features enable a person with a disability to interact with the contents in the smart mobile phones [15]. A large number of accessible built-in features have been developed specifically for people with blindness and visual impairment. Few examples are highlighted in **Table 1**. These features have been preserved mostly on Apple iOS and Android operating platform, contributing more than 90% of the world smartphone market [16]. The most widely used built-in accessible feature as screen readers are TalkBack for Android (**Figure 4**) and VoiceOver for Apple iOS (**Figure 5**) among people with visual impairment [17]. The Talback feature allows the user to easily identify the content or applications icons on the smartphones' screen with a verbally speak words or voice from the device upon touching with fingers on the icons. By simply placing a finger on the icons, the smartphone will read aloud what icon is

Features	Operating system	Operating mode	Descriptions
Voice assistant (Google assistant, Siri)	iOS/Android	Audio based	Ask it a question. Tell it to do things
TalkBack	Android	Audio based	Screen reader
VoiceOver	iOS	Audio based	Screen reader
Text to Speech/Voice recognition	iOS/Android	Audio based	Read aloud
Select to speak	iOS/Android	Audio based	Select to speak, Speak selection
Zoom magnification/Font size	iOS/Android	Visual based	Zoom in the display
Contrast	iOS/Android	Visual based	Differences between an object and its background.
Invert colors	iOS	Visual based	White becomes black, black becomes white, orange becomes blue.
Voice Inputs Keyboards	iOS/Android	Audio based	Typing through voice command

Table 1.
Accessible built-in features of smartphones.

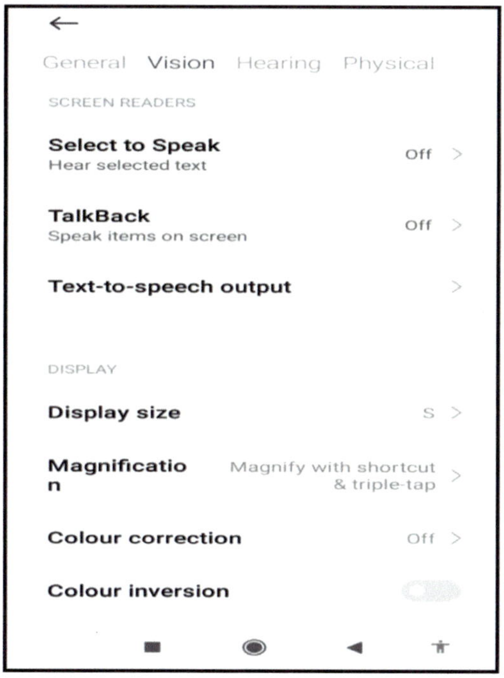

Figure 4.
Built-in features in Android operating system.

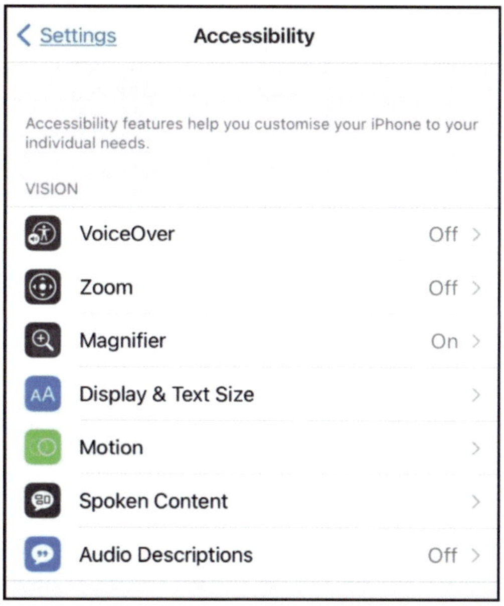

Figure 5.
Built-in features in iOS operating system.

underneath the finger. Similarly, VoiceOver provides voice feedback aloud what appear on the smartphones' screen, so iOS operating smartphone can be used without the need of visual function. A visually disabled individual can slide around on the screen with finger until the desired icons is located.

This feature provides unwavering support to perform activities of daily living, communication, social interaction, for a severe visually impaired or blind users, where human-vision smartphone interaction is impossible or difficult. For a person with a low vision, the zoom magnification feature helps to magnify the entire screen to the required level of individual's choice. Invert colors allows to change the white text on a black background, and vice versa or blue text on a yellow background A study on iPhone accessibility features in visually impaired population shows that the Zoom Magnification and Large Text were the most common use features among people with visual impairment [15]. This study reported that many participants were interested to use other accessibility features if they are familiar and trained on how to use such accessible features of a smartphones.

4. Third party accessible applications for visual impairment and blind

Generally, people believe that smartphones are inaccessible to people with a visual impairment; visually disabled people will face a lot of difficulties when they use for their everyday lives. This is because, operating smartphone is based on touch screen which is relied largely on good visual functions. When human-digital

Apps	Operating system	Operating mode	Descriptions
Kibo	Android	Audio based	Image reading, (pdf, e-book, doc, reader Hindi & English both
Be My Eyes	Android, iOS	Audio based	Help by Sighted person (volunteer) through video calling
Supersense	Android, iOS	Audio based	Information about surrounding, Artificial intelligence (AI)
Visor	Android, iOS	Video based	Magnifier (Near objects)
Binoculars	Android, iOS	Video based	Distance viewing (Super zoom camera)
Mani	Android, iOS	Audio based	Mobile Aided Note Identifier
BlindSquare	iOS	Audio based	Navigation, search places, etc.
Khabri	Android	Audio based	Audio news, current affair, job, horoscope, stories, and promotional podcasts Hindi, etc.
Phonepe or Google Pay	Android, iOS	Audio based	Easy and reliable way to make hassle-free payments online, E-Transaction
AccessNote	iOS	Audio based	Note taking
KNFB Reader	Android, iOS	Audio based	Its text-to-speech, text-to-Braille, and text highlighting tools make it valuable for blind
Blind Bargains	Android, iOS	Audio based	It brings you the latest sales, deals, and news on computers, screen readers, notetakers, Braille printers, hard drives, accessible cell phones, memory cards, talking products, household items, and much more
Seeing Eye GPS	iOS	Audio based	The Seeing Eye GPS, is a fully accessible turn-by-turn GPS iPhone app with all the normal navigation
Bard Mobile	Android, iOS	Audio based	It allows you to download and read audio books from the National Library Service (NLS) for the Blind, but needs to register in NLS
Audible	Android, iOS	Audio based	It provides a lot of audio books for people with visual impairment and blind

Table 2.
Some accessible applications for visual impairment of smartphones.

technology-interaction i.e., interface, is depend on vision, then visually impaired people face a tremendous limitation to access all kind of the information present in the smartphone. With ongoing advances and sophistication in mobile technology, the smartphone can be operated with eyes-free interface in an efficient way. These accessible applications are based on audio and tactile/haptic based interaction substituting the need of visual based interaction. There are many "apps" available that are accessible to people with visual disability. Few examples are given with a brief description of each in **Table 2**. **Figure 6** represents illustrations of few accessible apps. Majority of these apps are freely available online, so a visually impaired person can download, and install any specific app for their purpose.

These applications enable the smartphones serve as assistive technology for visually impaired people and facilitate to do tasks related to independent living, performing daily activities, to engage in education, societal activities, help in access information in today's digital society, sighted help, and to improve the overall quality of life. For example, the "Kibo" application helps an individual with a visual disability to access all electronic prints or pdf files or word document files (**Figure 7**); Be My Eyes for sighted help; Visor for color inverts; Mani app for currency identification and RightHear for navigation.

Figure 6.
Third Party accessible applications.

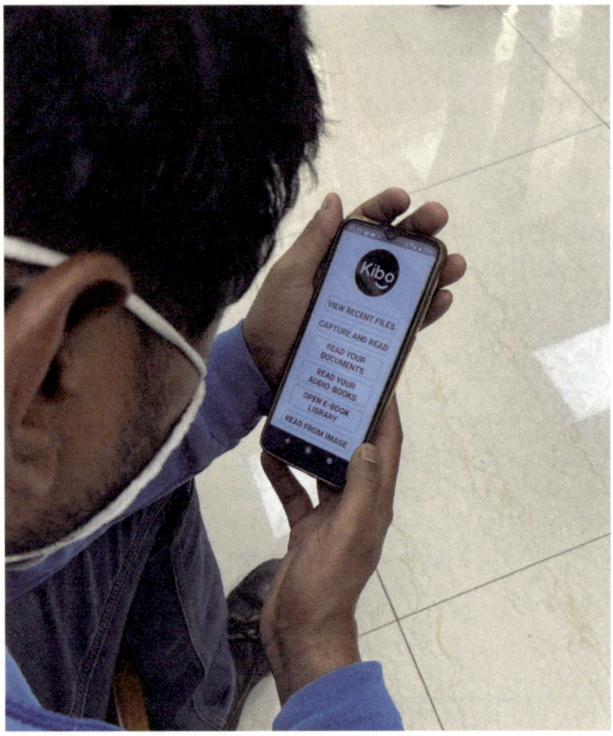

Figure 7.
Kibo accessible application.

5. Use among visually impaired and blindness

Smartphones, both its accessibility features and applications are being used for a wide range of daily living tasks that might have executed previously with the help of traditional assistive devices, e.g., magnifiers. The operating system can download a third-party accessible application to have a customised functionality. These accessible apps can be used for object and obstacle identification, sighted help, communications, emailing, reading e-Book and writing (Access Note, Be My Eyes, KNFB Reader, TalkBack, Braille Touch), news reading & listening (AccessWorld, Blind Bargains), entertainment, calendar functioning, currency identification, GPS navigation (Mani app, BlindSquare, Seeing Eye GPS) social networking, recording memos and color identification, talking calculator, so on.

An exploratory study on smartphone use among people with visual impairment shows that more than 90% of respondents used their mobile for activities such as calls, sending and receiving message, browsing the web, reading emails; 70–80% of them used smartphone for calendar functions, listening music and social media, and networking; 60 to 70% of them used for reminders and to take photos for reading with Optical Character Recognition [18]. This study also shows that 80% of participants used a smartphone for outdoor GPS navigation. An online survey conducted in a developed country in 2014 reported that 81% of respondents with a visual impairment used smartphone apps for various routine activities [19].

In a global survey on the use of accessible apps, more than 95% of people with visual impairment reported that these special apps were useful, and being used to accomplish their daily activities [20]. The most frequently use apps were Be My Eyes, ColorID, CamFind, followed by screening reading and writing apps, e.g. TalkBack,

KNFB Reader BrailleTouch. The study population of this global survey highlighted about the satisfaction of these special accessible apps of smartphones. In many cases, the smartphones apps can function for multipurpose tasks that facilitate independence living. Natalina et al. study also highlighted that the smartphones are replacing many traditional assistive solution to a great extend in doing daily activities [18]. A few studies reported that participating in digital arenas and accessing digital technology, including smartphones, reduce loneliness feeling, improve social contact, information sharing, gaining a better interaction with friends and family [21–23].

A study on tele-rehabilitation services for visually challenged students, in which smartphones were used exclusively, reported that smartphones based tele-health services offer a safe and an efficient way of providing all reliable information of COVID-19 pandemic, including various preventive strategies among students with visual disability. Such a tele- health services help to avoid the direct face to face contact between providers and patients. The study also highlighted such a platform helps in psychological counseling for fear and panic, facilitating and addressing the many unseen challenges faced by visually disabled people during the lockdown and pandemic period [24].

6. Advantages in using accessible features and applications

Conventionally, patients with low vision and blind receive optical and non-optical vision rehabilitation services using various traditional assistive devices such as magnifiers (both near & distance) and Braille, Digital Accessible Information System (DASIY) book and video magnifiers or Closed Circuit Television -CCTV (**Figure 8**). Although, the beneficial for traditional devices are widely documented in literature, their adoption and consistent use amongst visually impaired individuals is impeded by various factors, such as, discomforts, expensive, weight, difficult to carry, social stigma attached to it, lack of technical support in terms of training, lack of skills and knowledge in caregivers, unavailability [25–27]. For example, the cost of CCTV is 1000 USA dollars or more. Therefore, non-procurement by visually disabled is likely to be high. At the same time, the abandonment and disuse of these traditional devices are ranging from 30%, reaching upto 75% for few devices [28].

Mainstream smartphones are visually and physically demanding, and are well adopted by general population across the world, therefore when an individual with a visual loss uses a smartphone, the users are less likely drawn to public attention in comparison with traditional assistive devices. Furthermore, they are less likely to pose a social stigma while using them, less discomfort to use it and seldom have any

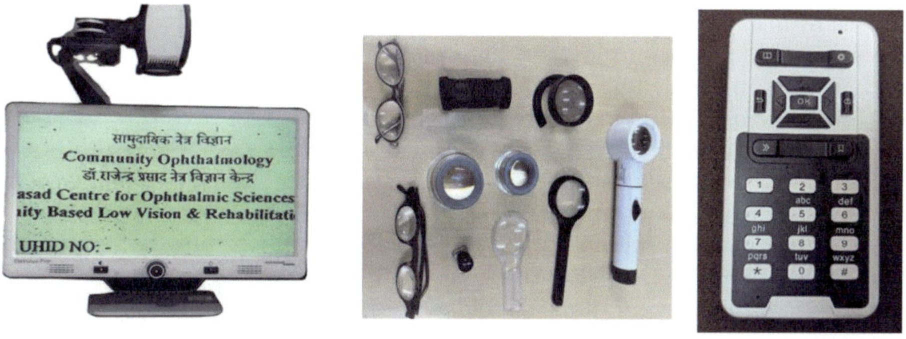

Figure 8.
Traditional assistive technologies (left to right: CCTV, Optical magnifiers, DAISY)

negative peer reaction on the users. As evidence exists on the influence of stigma and social negative reaction on adoption of tradition assistive devices in people with visual disability [18, 29].

In addition, the smartphones are available any time and any place, and universally design and user can carry everywhere [17]. The portability of smartphones with a lanyard or holsters is an important advantage as reiterated by visually impaired population. Their cost is relatively lower in prices compared to earlier periods. Therefore, the mainstream devices are replacing the traditional assistive devices in performing multiple daily living tasks, though the traditional devices are still in use for a limited specific purpose [18]. A study highlights that independent living, accomplishing daily living tasks will be better function in a person with visual impairment with app than without app [30].

The smartphones can be connected with the internet Wi-fi, Bluetooth for various purposes, further the accessible features and apps help to access various useful information. Such advance technologies help to accomplish a lot of activities from home avoiding the needless travel among the users. The new wireless charging system for smartphones is recently added on characteristics which can help to recharge the battery. With a new norm after COVID-19 pandemic, the smartphones and accessible apps help in availing tele-rehabilitation services provided by eye care centre [31, 32]. Such tele-health service help to avoid the unnecessary travel to the hospital and maintain social distancing during the lockdown.

7. Challenges in practices and application

As of today, many of us do not view smartphones as assistive technology for visual impairment. One of various reasons for it is, the healthcare providers, caregivers, even eye care professionals have a lack of awareness and understanding of the beneficial aspects of such smart mobile technologies, resulting in a lack of recommendation [33]. There are hardly any studies available about awareness on accessible apps among eye care providers or caregivers. Such studies are needed in the future while addressing the issues. Increasing accessibility and access or use of smartphones in persons with visual impairment requires sensitization and making understand the mainstream eye care providers, medical practitioners, including primary health physician, caregivers, family members at the fore front, even to a large section of the community. Therefore, digital literacy and providing information on accessible applications and features are required at all level health care delivery or hospitals, community, before generating awareness and skills development amongst beneficiaries. Accomplishing these activities should be one of the priorities of the vision rehabilitationists or low vision specialists, medical social workers, formal and informal community-based organization delivering services to visually disabled people. In this, the integration of assistive technology services for visual impairment, including smartphones into eye care practices or other related services for disabled could be worthwhile. There is a need for involvement of multiple sectors for such a large-scale activities, since involving only health sector is less likely to be a successful strategy. For example, educational sector for generating awareness among teachers, students; Information and Technology sector for enhancing learning, communication channels, labour market for easy availability, etc. should be involved in the awareness activities.

In addition, the relevant minitrial divisions of a country should have a policy to address the gap in the services of assistive technology for visual impairment, including smartphones and demands in their respective countries. International organization like The World Health Organization, Global Alliance for Assistive technology, UN convention for Rights for persons with Disabilities can support countries' policy and planning for assistive technology services, including service related to smartphones for disabled as a whole.

The number of accessible apps and accessible features are continuously growing over time for people with visual impairment. Designers are continuously fostering to develop more innovative and readily accessible, and user-friendly apps for visual impairment. These may lead to complexity on use of apps, since each app has a special function though some functions are overlapped. There is not a single app that fits all purposes. Therefore, initial assessment of the requirement for the apps will be required and followed by tailoring the training program to meet the specific needs. There is a need of developing training guidelines on use of such accessible apps or features that will facilitate to recommend appropriately for various types of functions and vision loss, e.g., reduced vision acuity, visual field loss, reduced contrast sensitivity or clinical findings. Using multiple apps in a smartphone may consume more power from the battery. So, there is a need of a good quality battery to operate extended periods of time, especially for smartphones used by visually impaired people. A due attention to improve awareness, on digital training, data security, back-up in the event of lost or theft of smartphones could be challenges among people with visual impairment.

8. Usability evaluation for visual impairment

People with visual impairment have a unique requirement to be able to interact with smartphone assistive technology. Such a unique requirement can cause a gap between the ideas of the developers of smartphones and the needs of individuals with visual impairment, leading to either a lack of adoption or abonnement of smartphone. There is an obvious need of collaboration between vision rehabilitation professionals and expertise in computer sciences and involvement of visually challenged individuals, and their participation, in designing an innovative, acceptable, and adaptable mobile assistive technology that will assist to develop a user-centred technology.

Therefore, usability testing or evaluation while developing of accessible features and apps for smartphones is essential so that the features and apps are accepted and adopted by the end users. For examples, a smartphone may have many built-in multiple functions, but if it may not be usable to target population due to inaccessible or discomfort on use. A systemic evaluation on the usability testing should be a part of developing for accessible features and applications of smartphones.

Usability is an elusive concept. To define it explicitly, we have to devise measurements that reflect the user's experience of a product and the level of success we establish for the product. As a working definition, usability is the ease with which people in a defined group can learn and use a product. The characteristics of and relationships among the ***tasks,*** the ***users,*** and the ***product*** determine the product's usability.

S. Rosenbaum Dec. 1989

The potential gap between the unique experiences and challenges faced by people with visual impairment and ideas of designers are crucially important. The designer needs to understand the user-centered and user-friendly perspective while efforts have put in to develop various apps as assistive technologies for visual impairment, that the developers create the apps that meet the user's expectation and needs. The product should be designed that a person with a visual loss can access all accessible features or functions that they want to execute. There is a need for evaluation and testing whether there are any usability problems in using the features and applications by end the users.

Designers or ergonomics engineers may face various challenges to perform usability evaluation for assistive products in an efficient way. Several factors may account for this, such as multiple influencing factors, so many non-specific tests and evaluation methods. There are methods available to do usability evaluation for assistive technology. Among them, Kwahk and Han has designed a simple usability evaluation framework of the electronic audio-visual products that can help the evaluators while conducting usability evaluation of their products in rational way (**Figure 9**). The framework is not specifically designed for smartphones apps and features, but the principles can be applied to other categories of assistive products with a minor change as per product designed. The framework provides a concise model how the evaluation should be done based on four parameters, i.e., first, user information; second, products information; third, user activity information; and fifth, the environment information [34].

The user information is important when the product is targeted at the users with special needs, for examples, individuals with visual disabilities or any other physical disability. The product information is required to understand the context of evaluation, for examples, form of the products- paper-based descriptions, mock-ups of the products or computer-based prototypes or finished products, etc.; level of the products in terms of perceived value: high-end, mid ranges, low end, etc. The environment and socio-cultural factors are also influence the user's operation and performance of the products.

Kwahk and Han further described five different techniques for evaluating usability of a product in human-electronic devices or computer- interface areas. The evaluator can select the suitable technique as per the context of the product and feasibility for conducting. They are as follows:

1. Observation/inquiry techniques

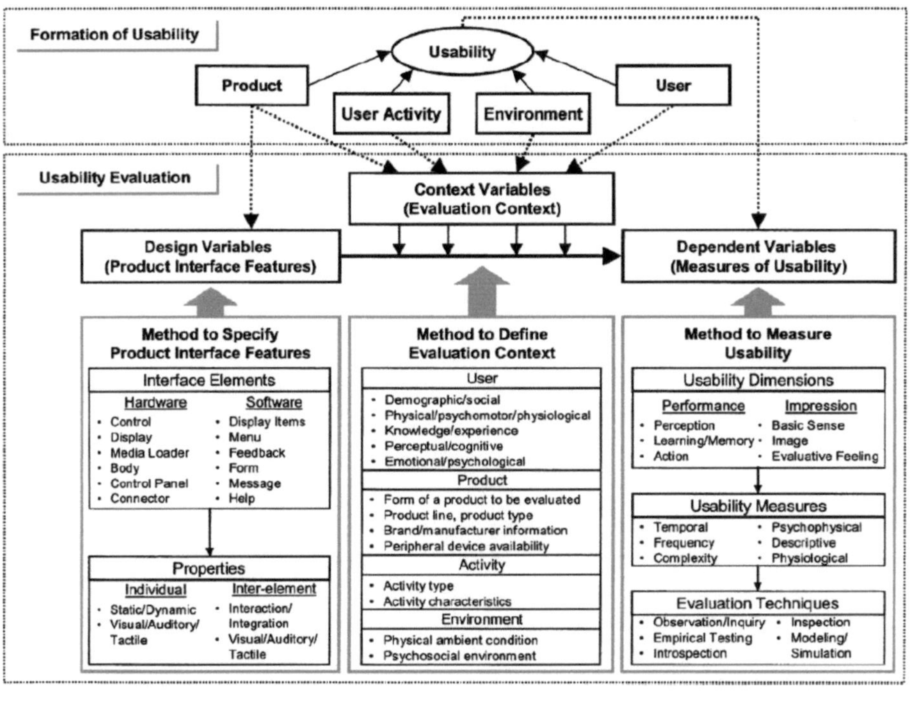

Figure 9.
Usability Evaluation Framework for electronic audio-visual products. (Source: Kwahk and Han [34]).

2. Empirical/usability testing

3. Introspection technique

4. Inspection methods

5. Modeling/simulation methods

The detail of each evaluation technique is beyond the scope of this book, but a brief description is given here. The empirical or usability testing with real users helps the designers to understand the experiences and challenges faced by people with disabilities while using assistive technology. Therefore, a designer can stress on empirical testing of the products [35]. Think-Aloud protocols are a dominant method in usability testing. There are two Think-Aloud techniques for usability testing that use commonly by the usability specialist; A Concurrent Think Aloud (CTA), in which experts request the participant to verbalise their thoughts as they perform the tasks with the help of the devices. Experts also use Retrospective Think Aloud (RTA), where participants are asked to retrace their steps after they complete the tasks and share the challenges [36, 37]. There are the pros and cons of both protocols, so expert has to decide which technique is suitable for usability testing.

Usability testing is the practice of testing how easy a design is to use with a group of representative users or persons with disabilities. It usually involves observing users as they attempt to complete tasks and can be done for different types of product designs. It is often conducted repeatedly, from early development until a product's release.
The goal of both usability evaluation and usability testing is the same: to improve the usability of products or assistive technology.

The introspection technique is an effective method to rule out the basic causes of usability problems based on user's thought and feeling. Often, evaluators might have difficulty to identify the cause of usability problems. This is because the evaluators have not imagined these problems that people with disabilities would have encountered while using the assistive products. Identifying the causes of usability problems helps the evaluators to differentiate whether the problems are due to individual's disability or by imperfect interface design for the assistive products [38]. For example, a study conducted by H K Kim et al. on identifying interaction experiences of visually impaired people when they use with smartphones, the common usability problems encountered are shown in **Table 3**. The study indicates that the problems vary according to the severities of visual impairment.

Usability evaluation	Usability testing
An ***expert evaluation*** is an assessment of the product's usability by an expert in usability issues, called a usability specialist. An expert evaluation, a usability specialist carries out an evaluation process that combines investigation of a product's specific user audiences and tasks with in-depth knowledge of general usability rules and measurements.	A ***usability test*** enables us to quantify the extent to which a product meets the needs of its intended users. Working with subjects selected as having the target users' point of view, the test consists of tasks designed to attain specified objectives consonant with the type of product and the characteristics of its users.

Usability problems
Performance of recognizing faces is lower than I expected
Difficulties of using the horizontal mode
Difficulties of finding the menu
Speech too slow
Color contrast
Difficulties in typing with a QWERTY virtual keyboard
Recording TalkBack sound along with my voice
Difficulties of understanding the location of a face on the screen
Correct pronunciation
Narrow touching area of menus
Screen reader voice unfriendly
Difficulties of selecting a shooting button
Beeping sound lasts too long
Difficulties of memorizing the interface layout
Need the function of voice pause
Need the function of voice stop
Difficulties of understanding the direction to move the camera
Locating similar menus too far
Menu buttons too small
Difficulties of inferring the word of big and small face
Speech too loud
Difficulties of using a sensitive touch interface
Need additional menu titles along with icon for first-time users
Too long photo title which is automatically inserted
Transparency of background
Font shape
Difficulties of recognizing the meaning of icon designs
Need more color combinations of text and background
Need bigger text size
Complex interface layout
Need informing the existence of sub-menus
Beeping sound pitched too high
Space between letters and lines
Stroke width
Need higher recording quality

Source: Kim et al. [38].

Table 3.
Usability problems identified in the study among visually impaired people.

9. Conclusion

Individuals with visual impairment and blindness face various challenges and barriers in their everyday lives. These people need assistive technology to overcome these challenges and perform a wide range of daily activities, including reading and writing. These technologies vary from low technology to high end and specialized technology, including software programs. In the recent times, mobile technology has gained a lot of attention across the world and are incorporated many advance computer and information technologies features.

Smartphones, the most sophisticated mobile technology, have been developed as smart assistive technology for people with visual impairment in the few years with many accessible built-in features and accessible applications with the help of sound, haptic and gestures interaction, instead of vision-smartphones interaction. Such innovative accessible, user-centred and friendly technology has provided a new platform and opportunities for people living with visual impairment to overcome the very challenges and barriers encountered in their everyday lives. Many accessible apps have been shown to be of great value and support in performing a wide range of daily living activities in people living with visual impairment. Such technologies are less likely attached with a stigma compared to traditional assistive devices. Further, use of smartphones have certain advantages, such as easily portable, less discomfort on use, relatively low cost compared to some of traditional devices. With widespread availability of mobile technologies along with eyes-free human smartphones features and applications, there might be a corresponding need of developing a clinical guideline on the use of accessible features and applications.

10. Recommendations

Now days, smartphones become a part and parcel of everyone lives, irrespective of health or disabilities. As of today, smartphones are asset of the general population who are sighted. However, with the help of accessible apps and built-in feature developed in the past few years, people with visual impairment and blindness can access smartphones and use for varied purposes. Now, it is time to scale up its use among visual impairment and blindness. Therefore, there is a strong need for sensitization, promotion of smartphones as assistive technology in health or eye care facilities across the globe, even in educational institutions. Integration of smartphones services as assistive technology for visual impairment, in all eye care practices, but not limited to vision rehabilitation, would be helpful in improving the accessibility of it. Considering the demand as well as requirement for smartphones among people with visual impairment, a strong and effective advocacy is the need of the hour, especially in low middle income countries.

Author details

Suraj Singh Senjam
Community Ophthalmology, Dr. Rajendra Prasad Centre for Ophthalmic Sciences, All India Institute of Medical Sciences, New Delhi, India

*Address all correspondence to: drsurajaiims@gmail.com

References

[1] Mathers C, Stevens G, Mahanani WR, et al. WHO methods and data sources for global burden of disease estimates 2000-2015. 2017 [cited 2017 Dec 1];1. Available from: http://www.who.int/gho/mortality_burden_disease/en/index.html

[2] Köberlein J, Beifus K, Schaffert C, Finger RP. The economic burden of visual impairment and blindness: A systematic review. BMJ Open [Internet] 2013 [cited 2020 Jul 9];3(11). Available from: /pmc/articles/PMC3822298/?report=abstract

[3] Senjam SS, Singh A. Addressing the health needs of people with disabilities in India. Indian J Public Health 2020;64 (1):79–82.

[4] Teutsch SM, Mccoy MA, Woodbury RB, Welp A. MAKING EYE HEALTH A POPULATION HEALTH IMPERATIVE VISION FOR TOMORROW. 2016 [cited 2020 Jan 3]; Available from: www.national-academies.org.

[5] Senjam S. Impact of COVID-19 pandemic on people living with visual disability. Indian J Ophthalmol [Internet] 2020 [cited 2020 Nov 9];68 (7):1367–70. Available from: /pmc/articles/PMC7574069/?report=abstract

[6] World Health Organization. World report on vision [Internet]. 2019 [cited 2020 Mar 26];Available from: https://www.who.int/publications-detail/world-report-on-vision

[7] Bhowmick A, Hazarika SM. An insight into assistive technology for the visually impaired and blind people: state-of-the-art and future trends. J. Multimodal User Interfaces2017;11(2): 149–72.

[8] Senjam S. Assistive Technology for People with Visual Loss. Delhi J Ophthalmol [Internet] 2019 [cited 2020 Nov 8];30(2):7–12. Available from: http://www.djo.org.in/articles/30/2/Assistive-Technology-for-People-with-Visual-Loss.html

[9] Hakobyan L, Lumsden J, O'Sullivan D, Bartlett *H. Mobile* assistive technologies for the visually impaired [Internet]. Surv. Ophthalmol.2013 [cited 2020 Jul 17];58(6):513–28. Available from: https://pubmed.ncbi.nlm.nih.gov/24054999/

[10] Martiniello N, Eisenbarth W, Lehane C, Johnson A, Wittich W. Exploring the use of smartphones and tablets among people with visual impairments: Are mainstream devices replacing the use of traditional visual aids? Assist Technol [Internet] 2019 [cited 2021 Jan 26];1–12. Available from: http://www.ncbi.nlm.nih.gov/pubmed/31697612

[11] Senjam SS, Foster A, Bascaran C. Assistive Technology for Visual Impairment and Trainers at Schools for the Blind in Delhi. Assist Technol [Internet] 2020 [cited 2020 Nov 9]; 10400435.2020.1839144. Available from: https://www.tandfonline.com/doi/full/10.1080/10400435.2020.1839144

[12] Product Narrative: Digital Assistive Technology | AT2030 Programme [Internet]. [cited 2021 Jan 31];Available from: https://at2030.org/product-narrative:-digital-assistive-technology/

[13] Singh SS. Smartphones as Assistive Technology for Visual Impairment. Eye 2021;in press.

[14] Android and iOS Combine for 92.3% of All Smartphone Operating System Shipments in the First Quarter While Windows Phone Leapfrogs BlackBerry, According to IDC | Business Wire [Internet]. [cited 2021 Jan 28]; Available from: https://www.busine

sswire.com/news/home/20130516005342/en/Android-and-iOS-Combine-for-92.3-of-All-Smartphone-Operating-System-Shipments-in-the-First-Quarter-While-Windows-Phone-Leapfrogs-BlackBerry-According-to-IDC

[15] Robinson JL, Braimah Avery V, Chun R, Pusateri G, Jay WM. Usage of Accessibility Options for the iPhone and iPad in a Visually Impaired Population. Semin Ophthalmol [Internet] 2017 [cited 2021 Jan 20];32(2):163–71. Available from: https://www.tandfonline.com/doi/abs/10.3109/08820538.2015.1045151

[16] Worldwide Quarterly Mobile Phone Tracker [Internet]. [cited 2021 Jan 28]; Available from: https://www.idc.com/tracker/showproductinfo.jsp?prod_id=37

[17] Irvine D, Zemke A, Pusateri G, Gerlach L, Chun R, Jay WM. Tablet and smartphone accessibility features in the low vision rehabilitation [Internet]. Neuro-Ophthalmology2014 [cited 2021 Jan 26];38(2):53–9. Available from: https://pubmed.ncbi.nlm.nih.gov/27928274/

[18] Martiniello N, Eisenbarth W, Lehane C, Johnson A, Wittich W. Exploring the use of smartphones and tablets among people with visual impairments: Are mainstream devices replacing the use of traditional visual aids? Assist Technol 2019;

[19] Crossland MD, Silva RS, Macedo AF. Smartphone, tablet computer and e-reader use by people with vision impairment. Ophthalmic Physiol Opt [Internet] 2014 [cited 2021 Jan 30];34(5):552–7. Available from: https://pubmed.ncbi.nlm.nih.gov/25070703/

[20] Griffin-Shirley N, Banda DR, Ajuwon PM, Cheon J, Lee J, Park HR, et al. A survey on the use of mobile applications for people who are visually impaired. J Vis Impair Blind 2017;111(4):307–23.

[21] Fuglerud KS, Tjøstheim I, Gunnarsson BR, Tollefsen M. Use of social media by people with visual impairments: Usage levels, attitudes and barriers [Internet]. In: Lecture Notes in Computer Science (including subseries Lecture Notes in Artificial Intelligence and Lecture Notes in Bioinformatics). Springer, Berlin, Heidelberg; 2012 [cited 2021 Jan 30]. page 565–72.Available from: https://link.springer.com/chapter/10.1007/978-3-642-31522-0_85

[22] Morten Tollefsen ØD. A paper about the disabled and the use of social media. [Internet]. 2011 [cited 2021 Jan 30]; Available from: http://www.medialt.no/pub/info_pdf/status_social_media_2010_english.pdf

[23] Fuglerud KS, Chan R, Sørli HT. Studying older people with visual impairments using mainstream smartphones with the aid of the Ezismart keypad and apps. In: Studies in Health Technology and Informatics. IOS Press; 2018. page 802–10.

[24] Senjam S, Manna S, Vashist P, Gupta V, Varughese S, Tandon R. Tele-rehabilitation for visually challenged students during COVID-19 pandemic: Lesson learned. Indian J Ophthalmol [Internet] 2021 [cited 2021 Feb 23];69(3):722. Available from: http://www.ijo.in/text.asp?2021/69/3/722/309386

[25] Phillips M, Proulx MJ. Social Interaction Without Vision: An Assessment of Assistive Technology for the Visually Impaired. Technol Innov 2019;20(1):85–93.

[26] Sugawara AT, Ramos VD, Alfieri FM, Battistella LR. Abandonment of assistive products: assessing abandonment levels and factors that impact on it. Disabil Rehabil Assist Technol [Internet] 2018 [cited 2021 Jan

28];13(7):716–23. Available from: https://www.tandfonline.com/doi/abs/10.1080/17483107.2018.1425748

[27] Pape TLB, Kim J, Weiner B. The shaping of individual meanings assigned to assistive technology: A review of personal factors. Disabil Rehabil [Internet] 2002 [cited 2021 Jan 28];24 (1–3):5–20. Available from: https://www.tandfonline.com/doi/abs/10.1080/09638280110066235

[28] Fuhrer MJ. Assistive technology outcomes research: Challenges met and yet unmet. Am J Phys Med Rehabil [Internet] 2001 [cited 2021 Feb 26];80 (7):528–35. Available from: https://pubmed.ncbi.nlm.nih.gov/11421522/

[29] Assistive Technologies for People with Diverse Abilities | Giulio Lancioni | Springer [Internet]. [cited 2021 Jan 30]; Available from: https://www.springer.com/gp/book/9781489980281

[30] Robinson JL, Braimah Avery V, Chun R, Pusateri G, Jay WM. Usage of Accessibility Options for the iPhone and iPad in a Visually Impaired Population. Semin Ophthalmol 2017;32(2):163–71.

[31] Senjam SS, Manna S, Vashist P, Gupta V. Tele-Rehabilitation for People with Visual Disabilities During COVID-19 Pandemic: Lesson Learned. medRxiv [Internet] 2021 [cited 2021 Jan 17]; 2020.12.31.20249111. Available from: https://www.medrxiv.org/content/10.1101/2020.12.31.20249111v1

[32] Senjam SS. A Persons-Centered Approach for Prevention of COVID-19 Disease and Its Impacts in Persons With Disabilities. Front Public Heal [Internet] 2021 [cited 2021 Feb 11];8:608958. Available from: https://www.frontiersin.org/articles/10.3389/fpubh.2020.608958/full

[33] Senjam SS, Foster A, Bascaran C, Vashist P. Awareness, utilization and barriers in accessing assistive technology among young patients attending a low vision rehabilitation clinic of a tertiary eye care centre in Delhi. Indian J Ophthalmol [Internet] 2019 [cited 2020 Nov 9];67(10):1548–54. Available from: /pmc/articles/PMC6786185/?report=abstract

[34] Kwahk J, Han SH. A methodology for evaluating the usability of audiovisual consumer electronic products. Appl Ergon 2002;33(5): 419–31.

[35] Rosenbaum S. Usability Evaluations Versus Usability Testing: When and Why? IEEE Trans Prof Commun 1989; 32(4):210–6.

[36] Van Den Haak MJ, De Jong MDT, Schellens PJ. Retrospective vs. concurrent think-aloud protocols: Testing the usability of an online library catalogue. Behav Inf Technol [Internet] 2003 [cited 2021 Feb 23];22(5):339–51. Available from: https://www.tandfonline.com/doi/abs/10.1080/0044929031000

[37] Weiyan Chee. Retrospective vs. concurrent think-aloud | by Weiyan Chee | Medium [Internet]. [cited 2021 Feb 23];Available from: https://medium.com/@cheeweiyan/retrospective-vs-concurrent-think-aloud-4e44e5d60143

[38] Kim HK, Han SH, Park J, Park J. The interaction experiences of visually impaired people with assistive technology: A case study of smartphones. Int J Ind Ergon 2016;55: 22–33.

Chapter 4

Usability Recommendations for Designers of Smartphone Applications for Older Adults: An Empirical Study

Bilal Ahmad, Ita Richardson and Sarah Beecham

Abstract

Older adults (OAs) are a growing and dominant part of the global population, with specific communication and usability needs. Information technology, such as smartphone applications, has the potential to help OAs stay connected, yet some designs do not appeal to this group of users. Current recommendations for the design of usable smartphone applications for OAs can be hard to apply and difficult to interpret. As a result, designers of smartphone applications do not have a clear set of recommendations for the design of smartphones for OAs. In this paper we elicit and transform usability trends and difficulties experienced directly by tech-savvy[1] OA users, into an organised set of recommendations. To do this we conducted an empirical study in four stages: (1) *Data extraction.* Digital context is extracted through conducting Think Aloud sessions with tech-savvy OAs (aged 50+); (2) *Data mapping*. Digital content extractions are mapped against 7 key aspects of usability; (3) *Validation*. Validated mappings through inter-rater reliability testing; (4) *Presentation*. Presented resultant recommendations as design patterns. Applying this method resulted in a set of 131 Usability recommendations with some overlap, transformed into a set of 14 design patterns that can act as a starting point for designers and developers of smartphone applications for OAs, and for pedagogy. Three of these patterns are presented in this study.

Keywords: Usability, Information Technology, Smartphone Applications, Software development, Recommendations, Design Patterns, Older Adults

1. Introduction

Across the globe, the proportion of older adults (OAs) is on the rise [1]. For example, Ireland's OA population increased by 19% in recent years [2]. This situation reflects that in most countries lifestyles have improved, however it does go hand in hand with impacting long-term care and welfare systems [3]. There are various definitions for what constitutes an older adult, and of course there is not a

[1] **Tech-savvy OAs** - OAs who spend more than 8 hours during a week on their mobile phone and use basic functionality as well as mobile applications and wearables. Examples of mobile applications are WhatsApp, Podcast, Headspace, and wearables such as Fitbit.

particular day in which someone moves from being a younger or middle aged adult to an older adult [4]. For the purpose of our study, however, we define OAs as someone who has reached the age of 50, basing our work on established studies e.g. TILDA[2] [5]. Additional qualifications to be included in our study is that the OA is already motivated to use smartphone technology.

The quality of life of OAs can be considerably improved through active social engagement, improved healthcare, and increased mobility through access to transportation [6]. Technology offers a ray of hope in providing access to these opportunities. But, the adoption rate of technology by OAs appears low [7]. Technology comes in many forms, to include public displays, virtual reality, websites and smartphone applications. As numerous technologies may have a different set of design requirements, our focus will be on smartphone applications only – smartphones are frequently used by OAs [8]. Furthermore, smartphone applications can help to provide access to digital services with no or low-cost [9].

To develop smartphone applications for OAs, designers must recognise the individual needs of this group such as the varying degrees of physical and cognitive decline [10, 11] and privacy concerns [12]. These specific features are not always considered, and a recent study advocates for better recommendations for age-friendly design of user interfaces on mobile phones [13]. A tried and tested way to elicit such specific recommendations is through participatory or human-centred development [14]; shown as a successful way to develop such guidelines [15] and checklists [16, 17], and for interface design of mobile phones for older adults [18]. However, existing guidelines deal with visual and haptic issues and omit many elements associated with the textual interface. Additional shortcomings of these guidelines are that they are rarely tested [15, 18] and lack extensive empirical validation [19]. Also, the guidelines for designing applications for OAs are not presented to designers and developers in an accessible format [20] and are sometimes merged in the discussion section of the papers e.g., [19, 21, 22]. A recent literature review summed up existing guidelines as piecemeal, lacking in characterisation, not easily actionable, and rarely validated [23]. In summary, the conflicting and disparate advice offered, results in arguably unreliable and hard to follow guidelines.

In this chapter, we focus particularly on the Usability recommendations for smartphone applications, for designing products to be effective, efficient, and satisfying. Usability includes user experience design. In our context, usability covers general aspects that impact everyone and do not dis-proportionally impact people with disabilities [24]. Furthermore, we look for better ways to disseminate our guidelines by transferring the recommendations into a set of patterns. This form of reporting is shown to lead to increased adoption of guidelines by practitioners [23, 25, 26].

We respond to the call to create a design patterns library aimed towards practitioners and pedagogical settings [20].

In summary, a few attempts have been made by the research community to produce guidelines or checklists, but for the most part they have not been validated or applied in practice [27]. Also, their implicit nature lacks explanation on how to apply these recommendations. Henceforth, this research fills this void by providing a structured and validated set of design patterns to support the design and development of usable smartphone applications for tech-savvy OAs. Our study aims to answer:

Research Question - What do tech-savvy older adults expect from smartphone applications?

[2] The Irish Longitudinal Study on Ageing, conducted at Trinity College: Dublin - URL: https://tilda.tcd.ie/

The next sections of this study are organised as follows. Section 2 describes the four different stages within the Research Methods and how the collected data was analysed. After that, Section 3 presents the results of execution of these stages and Section 4 depicts the transformation of findings i.e. recommendations in the form of design patterns. Section 5 then highlights how some of the threats to validity were catered for whilst conducting this research. Section 6 concludes this study and points towards potential future directions.

2. Methodology

2.1 Data collection methods

Data collection was conducted in two phases, firstly through Think Aloud sessions, followed by Digital Content extraction.

2.1.1 Think aloud protocol

The primary objective of the first phase i.e., Think Aloud (TA) sessions, was to understand the expectations of tech-savvy older adults regarding a set of smartphone applications. Think Aloud Protocol is a tool borrowed from cognitive science [28]. There are numerous variations of this method for Usability evaluation of technology. However, there is no consensus about the exact procedure to perform these sessions [29, 30]. These are usually performed by asking individuals to think out loud whilst interacting with the system [31]. If the user stops talking for long, then a gentle reminder is used to prompt him/her to start conversing again. The other option is to have a pair of users verbalise their thoughts retrospectively rather than concurrently [32]. For this study, the Think Aloud Process of [33] was used for conducting the sessions (as described below), because it is well established, widely used, and helps in Usability testing as shown in **Figure 1**. OAs in our study were given the liberty to choose two smartphone applications (already installed on their smartphones), use them, and highlight the good and bad things about them. Author 1 observed each OA and took field notes as they navigated through their chosen apps.

Instructions and Tasks - Older adults were instructed about how to interact with their mobile phones and given an explanation of how to think out loud. In addition, a pre-specified set of tasks was given to them in advance.

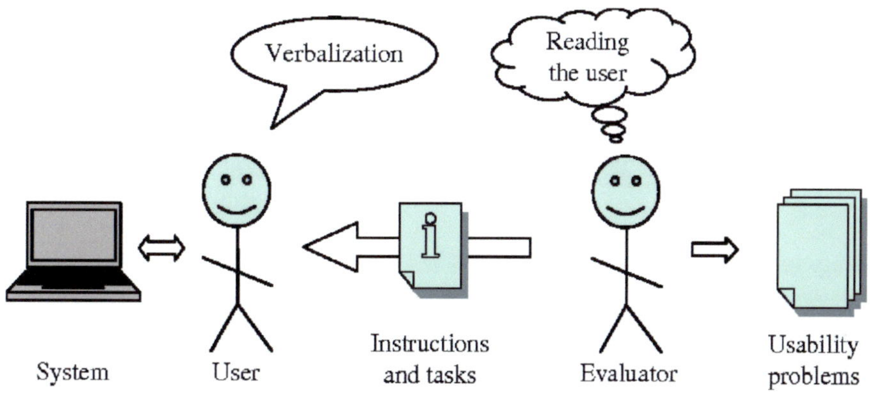

Figure 1.
Think aloud process [33].

This means that they had to use applications of their own choice and use one or two key features on them.

Verbalisation - During the process of performing the tasks, participants verbalised their thoughts. If they stopped talking, a gentle nudge was used to prompt them, e.g., through asking a question or just saying 'keep talking'.

Reading the User - Older adults' behaviour was observed to extract and describe the Usability problems. For example, how do they navigate the app? What elates or annoys them? What causes deadlock?

Relationship between User and the Evaluator - The researcher attempted to create an environment in which older adults felt comfortable to make both positive and negative comments. To achieve this, the venue chosen was a Cafe situated in the University of Limerick.

2.1.2 Digital content extraction

The second phase to further understand tech-savvy OA's user-interface needs was digital content extraction that comprised two sub-phases:

1. **Ageing Forums** - In order to strengthen the findings about the needs and wants of older adults who are tech-savvy, data from online ageing forums was collected. The recent trend of qualitative Internet research has resulted in a set of guidelines to effectively conduct such studies e.g., [34] and we have used the guidelines of [35] due to their practical nature. Two online forums for the 50+ age group were chosen, Seniorsonly[3] and SeniorForums[4]. The rationale for choosing these two forums was that they were both established forums (active since 2020), aimed at over-50 year olds, with active question and answer sessions and have a total of 4,500 subscribers. Recent conversations were observed. Participation in these groups as part of this from research was done in a non-invasive fashion.

2. **Mainstream Apps for Older Adults** - We investigated apps for older adults which were available on the Apple iTunes and Google Play store, as these two platforms share a combined market distribution of 99% of worldwide smartphone platforms as of December 2020[5]. We extracted the description and qualitative feedback pertaining to these apps. The objective of this data collection was to identify what older adults like and dislike and any changes they expect in the apps developed for them. The selection process of the applications for older adults and associated details is provided in [36]. Initially, the title, description and URL of the apps was extracted. Author 1 installed these apps on both an iPhone 6 and an Android phone and extracted the reviews, comments, and responses available on the play stores. The apps were then used for 10–15 minutes approximately to finally extract the key aspects, functional and non-functional, of the app. The qualitative data was accumulated and merged into an excel sheet.

2.1.3 Map recommendations

We coded and classified the recommendations into a structured vocabulary using a deductive approach. The recommendations were mapped into seven

---

[3] http://www.seniorsonly.club/

[4] https://www.seniorforums.com/

[5] https://gs.statcounter.com/os-market-share/mobile/worldwide

categories provided in Peter Morville's [37]. Usability honeycomb as shown in **Figure 2**. Usability is about designing products to be effective, efficient, and satisfying. Usability includes user experience design. This may include general aspects that impact everyone and do not dis-proportionally impact people with disabilities. The caveat here is that usability practice and research often does not sufficiently address the needs of people with disabilities [24].

We also considered two alternatives to Morville's definition of Usability. Firstly, in 1994, Jakob Nielsen [38] provided 5 attributes that together constitute usability i.e. Learnability, Efficiency, Memorability, Errors and Satisfaction. Then in 2003, Whitney Quesenbery [39] proposed a 5E model to describe usability i.e. Effective, Efficient, Engaging, Error Tolerant and Easy to Learn. Subsequently, in 2004, Peter Morville used a 'honeycomb' to illustrate usability with 7 sub-categories Useful, Usable, Findable, Valuable, Desirable, Accessible, Credible. The reason we choose the latter is that it is the latest and more granular than the former ones i.e. 7 categories compared to 5, and it is well recognised and highly cited.

2.1.4 Validate recommendations using inter-rater reliability

Different coefficients can be used for evaluating the agreement in classification of recommendations between the three raters or inspectors.

Proportion Agreement- A straightforward approach to evaluate the agreement is to consider the proportion of ratings upon which raters agree. This is, however, considered naive as the agreement may have occurred solely by chance. According to [40], using proportion or percentage of agreement tends to produce higher values than other measures of agreement. He discourages the use of proportion agreement, because science is inherently about conservatism rather than liberalism. In addition, the use of the proportion of agreement can be unreliable [41]. Therefore, the use of proportion or percentage agreement was not our choice for an evaluative measure.

S-Coefficient- Another option for evaluation of the agreements was the S-coefficient proposed by [42]. However, he assumes that the agreement by chance is due to raters assigning sub-categories/classes to the recommendations randomly at an equal rate.

Cohen Kappa- An alternative definition for agreement is the raters' tendency to distribute the classifications in a certain way. This seems a reasonable assumption

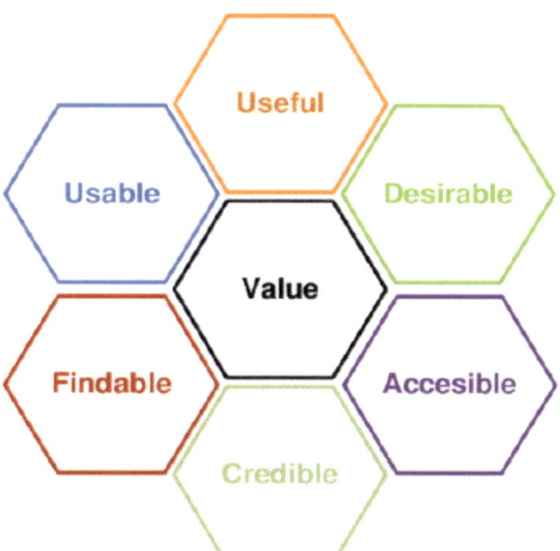

Figure 2.
User experience honeycomb [37].

a priori, in an inspection context. This is assumed to be the case with Cohen Kappa's coefficient [43]. We chose it because the three researchers, based on their theoretical knowledge of the domain, would be expected to classify the recommendations in a specific way, and given that we have 7 options in which to classify the text snippets from the forums, there was plenty of room for error or to highlight differences. It has been established that a good agreement as measured by Cohen Kappa's coefficient can produce slightly higher reliability results in the case of over seven categories [44]. Moreover, the Kappa coefficient is widely used in social and medical sciences and it has thousands of citations to date [45]. In the medical domain, it has been presented as a measure of agreement in reliability studies [46]. A variant of Kappa called weighted Kappa [47] was also considered, but it is most useful for non-nominal scales and when the relative costs of agreement can be quantified. The analysis of these three options led us to use Cohen's Kappa [48] because it is a robust and useful statistic tool for inter-rater reliability testing.

2.1.5 Transform recommendations to design patterns

Once we identified and developed a consolidated set of recommendations, we structured them into a design pattern format which consisted of sections: Problem, Rationale, Solution, Type, Sub-type, Related Patterns, References/Evidence as shown in [36], which explains each heading and what is included in each section.

Qualitative evidence suggests that design patterns can aid maintenance, but they do not appear to help novice designers learn about design [49]. Also, Bieman particularly observes that: 'there is very little empirical evidence of the claimed benefits of design patterns and other design practices when applied to real development projects' [50]. This implies that simply using patterns does not ensure good design, they should be used appropriately [49]. The initial attempt to develop design patterns was made by Christopher Alexander to document and discuss building architecture design and solutions [51]. Design patterns have become popular among software developers after the publication of "Gang of Four" [52]. It is defined as:

"A pattern is a structured document comprising of set of pre-defined sections; this means that all patterns in a given pattern language have the same general structure, making it easy for readers to find information" [53].

There are subtle differences in the structure of the design patterns used in existing research. This paper will follow the format proposed by [54] as this is also used in numerous studies e.g., [53, 55, 56].

2.2 Data analysis strategy

In qualitative research, the process of data collection, data analysis, and report writing is not always completed in distinct steps; they are often interrelated and occur simultaneously throughout the research process [57]. We used thematic analysis to develop themes that were emphasised during Think Aloud sessions or on the ageing forums. This is a method for identifying, analysing, organising, describing, and reporting themes found within a data set. Bruan et al. 2006 [58] presents a linear, six-phased method for thematic analysis. This involves becoming familiar with the data, generating initial codes, searching for themes, reviewing themes, defining and naming themes and producing the report. It is actually an iterative and reflective process that develops over time and involves a constant moving back and forward between phases. We have adapted the guidelines of [57, 58] to perform analysis of the qualitative data generated during the mixed methods. This was accomplished using Microsoft Excel. The snippets were matched with the theme columns and marked or annotated using zero, one, tick, cross. To facilitate code creation, we used a single 'sentence' as a unit of analysis for extracting and defining the themes. Author 1 developed an initial set of themes which were validated by author 3.

Step 1: Reading- Author 1 browsed through the transcripts verbatim and took notes from first impressions. After that, he re-read the transcripts line by line.

Step 2: Coding- Phrases, sentences, and sections with regard to the importance were labelled. A sentence was considered as important or relevant based on several factors, e.g., if the participant has explicitly said that this aspect is important, whether something new or surprising came up, whether something that had been previously published was discussed, or if something is repeated in several places.

Step 3: Categorising- Analysing the all the codes created, the most important and relevant codes were combined to form a category or a theme. Some trivial codes which had no relevance to the research objectives were ignored. The themes that were generated as a result of this stage were general and abstract.

Step 4: Label and Define the Categories- These high level themes were labelled and developed from the significant codes. These high level themes were the main results of this study which elaborate new knowledge from the perspective of tech-savvy older adults. The definitions of these themes were also outlined.

Step 5: Rectification of Unseen Bias- In order to alleviate any bias in the findings and to increase reliability, Author 1 randomly sent the snippets to author 3 without a theme assigned. The list and definitions of themes was sent separately. Author 3 commented, assigned these snippets to the themes and identified new themes. This resulted in the addition, deletion and modification of several themes. Any unresolved issues were discussed between all authors until a consensus was reached. After the completion of this step, we had a final set of mutually agreed 3 key themes and 24 sub-themes.

3. Results

3.1 Think aloud and digital content extraction

3.1.1 Think aloud protocol

Ten OAs, five female and five male, who had English as their primary language, participated in Think Aloud sessions. All of them had access to basic technology such as landline, internet and smartphones. Their median age was 64.7. Eight were fully or partially retired and two were working professionals. OAs were delighted to talk about the apps they use on a regular basis. For instance, the first participant, a retired Civil Engineer, talked about his experience with SmartTools[6] app. He said that SmartTools is his favourite app because he can perform basic civil engineering tasks using this simple and easy to use interface showing the tools. The tools include ruler, protractor, level, thread, range-finder, compass, metal detector, sound meter, vibrometer, flashlight, magnifier, mirror and unit converter. He was pleased about the real-world metaphor of 'Cupboard' used in this app. Also, he said that the labels used in the app are self-explanatory. On the other hand, he did not like using a budget airline app. He said the initial waiting time, excessive number of advertisements, hectic process to book a flight, small text size, icons without labels, and long and technical terms and conditions make him reluctant to use this app. He stated that this app does not cater for OAs and gave an example of a issue in the form completion section. The salutations, Mr, Mrs, need to be tapped for successful

[6] **Smart Tools:** http://androidboy1.blogspot.com/2015/12/smart-tools-v20.html

submission of the form, but there is no indication to select them. Also, they are represented in dull colours, reducing their importance in the eyes of the user. An array of recommendations given by this participant was collected such as "video based tutorials in the app to enable the user to become familiar with the app". Similarly, the analysis of data collected from other participants helped to generate recommendations.

3.1.2 Digital content extraction

We extracted data in two ways, firstly from online forums in which OAs were engaging with developers and the wider community on issues they were experiencing with their apps (see (A) below), and secondly by testing a sub-set of apps developed especially for OAs (see (B) below).

A. **Ageing Forums** In order to strengthen the findings about the needs and wants of OAs who are tech-savvy, data from two online ageing forums was collected.

1. **The Seniorsonly[7] Club** - run by seniors for seniors. There are eight different sections of this forum on which seniors can post questions, Forum Administration and Notices, Education and Entertainment, Getting it Done, Hot Button Issues, I remember when, On the lighter side, Ourselves and Others, Off topic. In addition, the technology section on this forum was active and recent conversations were observed. Author 1 delved deeply into a subsection of 'Education and Entertainment' called 'Gadgets and Tech talk'. The objective of this section was to discuss seniors' issues related to electronic or mechanical gadgets, equipment, computers or computer software, phones or smart phone applications, electronic reading devices, or anything relating to gadgets and technology. Fifteen threads with a total of 238 posts were included during data collection of this study.

2. **SeniorForums[8]** - Author 1 also looked into the 'Computers and Phones' section of SeniorForums, another forum for people over 50. Recent conversations in the technology section on this forum were observed. A total of 15 threads related to technology, particularly mobile phones, consisting of 170 posts, were included in this study.

OAs on these forums appear to be well-versed with technology. They would love to use more of the latest technology if it is inexpensive and privacy is ensured. This is in line with our previous research [59]. One of the older adults mentioned on the forum that he would prefer wearables rather than smartphones. Here is an excerpt from his comment:

"When they make them so that the Apple Watch is my cellphone, then I might consider it because carrying my iPhone is never convenient for me. I don't like carrying a phone in my shirt pocket, although having had cancer twice now, I don't suppose that's something I should worry much about, and there's no other convenient place to put it. I keep leaving it on the table at restaurants and have to go back for it. I would wear a watch if I could use it as a phone."

---

[7] http://www.seniorsonly.club/

[8] https://www.seniorforums.com/

OAs constantly discuss how to ensure privacy whilst using technology on these forums. Here is a solution proposed by a member on the question of ensuring privacy when using websites:

"Being stuck with a fixed, constant ISP number hinders one's privacy quite a bit. A roaming ISP number helps. Using a Proxy service server is useless, as I've found out. If you really want to operate some of your P.C. activities, do it at a public computer place, library, or the like, while setting your PCs security settings at maximum, reject all cookies. Set Temporary Internet Files to store for 0 days, and delete automatically each time you log off. Check them now and then for cookies, anyway. Always use In Private (Internet Explorer) or Incognito (Chrome) to boot."

B. Mainstream Apps for Older Adults.

A total of 32 mainstream apps were evaluated to find recommendations that are making these apps successful and/or unsuccessful. We started by identifying keywords, joining them together to define search strings. An example search string applied to app stores is 'apps for older adults'. Author 1 then searched Android and Apple app stores using the defined search strings. An initial search produced 177 and 18 results for the former and latter (these searches were conducted between May to August of 2019). A scrutiny was performed on these retrieved apps using an agreed inclusion and exclusion criteria [36]. This exercise reduced the number of apps, and we were left with a set of 101 android and 6 iOS apps. As the evaluation of 107 apps would have required a herculean effort, and more time than we had available, we selected a sample of apps from this set. In order to reduce the sampling error and achieve precision, we opted for a stratified sampling method. Hence, these 107 selected apps were carefully reviewed, arranged and classified into 9 different categories. These were Health & Safety, Launcher, Dating, Gaming, Shopping, Entertainment, Communication & Socialisation, Education and Tools. After stratification of the apps, a simple random sampling technique was applied on each of the nine sub groups. An equivalent percentage of 30% was selected from each type of app. This step resulted in a total of 32 finally selected apps, 30 android and 2 iOS based.

3.1.3 Qualitative analysis results

We analysed the qualitative data generated from Think Aloud Protocol and Digital Content Extraction using the strategy explained in Section 2. Likes, dislikes and recommendations for the development of smartphone applications for tech-savvy older adults were extracted.

1. Likes.

Source of Information Older adults really like a feature that serves as a source of information.

"This is a comprehensive app, which gives much information that the elderly need to know, from taking care of their health to advice on how to handle their finances to how to secure their property."

"Every information is available including elder abuse and rights of the elderly."

"It is a user-friendly app to get any sort of information on elders."

Video Tutorials Older adults liked to learn new information or even exercises through video tutorials available within the apps for them. For example, one of the older adults commented:

"Tried a lot of apps, finally landed here, it is a simple and useful app. The best part is the video tutorials".

Support Independent Living In addition, an app that supports independent living was really admired by older adults. One of the comments on the app stores was:

"My mom was looking for a personal fitness coach. Apparently, this app helped her achieve fitness goals without a trainer."

Allow Customisation Older adults preferred apps that allow customisation according to their needs and wants.

"After trying several app launchers, I settled for this launcher. Simple to use with enough customizing ability. Very clean too."

"Good app and easy to customise. Only if I could add widget and one more screen tab. It will be 5 star app."

Easy to Use and Learn One of the key things older adults look for in an app is how easy it is to use and learn.

"Recently, gifted a 5.5" android GS370 to my 70 year old mother and this launcher was one of the reasons to do so. Requires little to no explanation, easy to learn and manipulate - especially for the elderly. No superfluous apps installed or subscriptions required. Really appreciate the efforts by the developers."

"Great launcher, simple to use. I installed it for my mother. She is bad with technology, but got the hang of it pretty easy. 5 stars is not enough."

"Oh my God! so easy to use for seniors. No ads."

"Good for my mother, she feels it is easy to use."

Big and Attractive Icons Big icons also make it easier for older adults to operate the application and they mention it as often as they could at any forum where they had the liberty to speak about their expectations from an app for them.

"I like the big icon."

"My mom's phone also having it now. Big icon easy to operate."

"Great user interface, user friendly look like a stock launcher. Icon is bigger than others nice launcher."

If icons are ugly, older adults get demotivated and do not want to use the application or a particular version where the icons are not attractive.

"Downgraded to 4 stars due to ugly icons in latest update. Why? Actual bugs remain unfixed."

2. Dislikes.

Non-functional Features If a feature in an app did not function the way it was intended, then it reduced the motivation to use the app for older adults and they became sceptical of the entire system.

"Nobody receives the SOS calls."

"SOS is just an illusion created to attract people. No actual help from their side whatsoever."

Many Advertisements Older adults could only be truly satisfied with an app if there were no distractions, e.g., advertisements. One of the users clearly mentioned that he/she would only give a high rating to the app if there were not advertisements.

"Not downloading and only ads after each minute, fix this and I will give 5 stars".

Lack of Explanation Older adults dislike an app which did not clearly explain how to use its features. An example was a fitness app which tells the user what to do, but did not explain how to do it neither in descriptive nor in pictorial or video format.

"Not much to it ....tells you how many reps to do but doesn't show or explain the exercise."

Non-durable Features Older adults do not like a non-durable application which has no reliability as to whether it will work or not. This means that if it works for some time and stops working or freezes or crashes during use, this will lead to un-installation of the app by older adults.

"Worked well for 6 days then packed in and refused to start even after uninstalling and reinstall."

"Crashes immediately when swiping up to open app list. I tested 2 different Huwaei Phones. Honor 7X and Honor 8. Both crashed."

"Since last update, today, has crashed at least 17 times in 30 minutes. Fix it".

"Had potential. Updates broke it."

"Excellent app, easy and nice large lettering, the developer of this app did a great job, it certainly deserves 5 STARS, have re-downloaded and working again fine, but still freezing after phone shut off or changing to another launcher temporarily."

"Crashes constantly."

Access to Sensitive Information OAs are very sceptical and want to understand the reason why an app would be constantly asking for access to files and location. They are very concerned about privacy and confidentiality. This is one of the consistent findings of our research, which is contrary to the widespread perception that OAs do not care about privacy and are easily scammed.

"Mandatory weather widget that demands location data. Weather widget won't allow users to enter their city manually, these developers want the user's exact location."

"The app requires a lot of permissions and refuses to even start without them. For example, must have access to your location."

"Access your files and know your location, Why?"

Compatibility Issues OAs want the app to work on any type of device they own e.g., android, iOS or windows based. They also want the app to work perfectly on different sizes of device, whether large or small.

"Not working on my Moto G6. From the home screen, every time I swipe up to open the application drawer Simple Launcher crashes. I want to use this app, but can't. The app does work on my OnePlus 6 without any problems, but crashes on my Moto G6."

"Please add support for long screens. In Redmi 5 I can see black bar on top."

Confusing If it is confusing for OAs to find or use the application, they become reluctant to adopt it. Here is one excerpt from a set of OA comments on an app designed for them.

"How to use it man..I am not even finding...where it had gone after my download.. Please find it for me."

3. Recommendations.

Constant Improvement To maximise usage, the apps must keep on evolving and improving with the passage of time, even if the users, older adults, are happy with the current version.

"Installed this app on my fathers phone. The exercises are good and basic. Looking forward to a better experience."

Pick and Move In case of an app that falls under the category of a launcher, OAs want to move the apps freely on the interface, rather than deleting and relying on auto-arrange. This recommendation is about those launchers, which currently do not allow drag and drop to move apps.

"Can you make it so the apps can be arranged by moving them around. The only way I can figure out how to do it is to delete them and carefully plan which apps, I would like in order. Not sure how many stars to rate the app as yet."

One Tap Calling The app should allow quick calling to the contacts through a single tap, instead of conventional navigation which requires multiple steps.

"Just enough! Great for seniors! But, I would love it more if it gave the option for one-touch calling. Currently, contact icons bring you to the contact page. Please enable a direct dial option, as this would make the app perfect!"

"Please try to dial contacts directly with one touch."

Multiple Themes The app should have a feature that allows you to choose a theme, colour combination, of users own choice.

"It would be nice to apply different icon packs for a really cool look."

"Nice simple launcher but would like to see more features like themes and icon pack support also ability to add widgets."

Allow Remove Feature The app should allow removal of any features which older adults do not want to use any more. This will also make the interface cleaner and simpler.

"Ability to remove SOS for those who do not need it and make the page a full screen for contact shortcuts."

Notification Badge on App Icon Older adults like to have notifications badge on the icon of the application to reduce annoyance.

"Really great app, but I'd like to have the option to display the date in UK format (dd/mm) and to change the font size and colour. It also doesn't show notifications on each app's icon - is this something the launcher could do?"

"Add a badge notification option for apps. I never know how many texts I have until I open the app which is annoying."

Incorporate Hold Down Gesture In some launchers, hold down gesture does not work, which older adults would love to use.

"Good clean nice looking launcher with a few missing things. 1: Holding down on an icon for a while does not open it. 2: On the contacts screen there is no option to SMS or send a message to the person. While I feel that it is good for many people to hide this option, for others it is essential."

Disable Hamburger or Drawer Icon Older adults can accidentally tap the drawer/ hamburger which opens the detailed menu making it confusing for them and difficult to return. It is therefore recommended to disable it in the apps for older adults.

"Please add an entry in the options menu to disable the app drawer so that an oldie does not swipe up and show the drawer by accident. This way the usable apps are limited to what is shown on the main launcher screen. Great app and great support. Thank you!! "".

Improve Colour Combination The apps for older adults must cater for users with visual problems or those who are colour blind. This means a high contrast should be used, which is a recurring theme that emerged throughout the previous studies as well as the evaluation of apps.

"Why did you ruin a great app. The notification panel has become transparent which looks odd over the dark launcher screen...rollback the update or fix it ASAP please."

Allow Multiple Metrics The features in the app must consider different locations and cultures and incorporate multiple metrics for easier understanding, particularly for older adults.

"Locked in metric. Can't change to Fahrenheit."

"So far really nice... Big icons and super easy interface..only complaint is that not sure how to set temp to Fahrenheit for us non metric luddites."

"Perfect except needs Fahrenheit not just Celsius."

Make Swiping Fast Older adults get concerned if the apps execution is slow, in particular with regard to swiping. They admire applications which have fast swiping.

"Great app, but please make the swipe up fast."

3.2 Map and validate recommendations

To reiterate, the Think Aloud Protocol and Digital Content Extraction were conducted concurrently. This produced an exhaustive list of recommendations to

Recommendation Description	Think Aloud	Digital Content Extraction
Incorporate generous spacing between the buttons in the application.	✓	
Label the icons in the application.	✓	
Make the app accessible without the need for a password.	(✓)	(✓)
Avoid the advertisements in the application.	✓	✓
Incorporate a home button in the app serving as a safe point of return for older adults.	✓	
Use minimal parameters for the request to access personal information in the installation prompt whilst installing the application.	✓	
Avoid technical jargon in the application.	✓	✓
Allow the older adults to choose a preferred theme in the application.	(✓)	(✓)
Include features which are necessary in the application only.	✓	
Enlarge the size of the components used in the application.	✓	✓
Allow performing most functions in the application through tapping rather than drag and drop.	✓	
Avoid the use of new and unfamiliar gestures in the application.	✓	✓
Use culturally acceptable terminology and icons in the application.	✓	✓
Use layman's language in the application for error notifications.	✓	
Incorporate a tutorial in the application to teach older adults how to use the application.	✓	
Use precise and easy to understand terms and conditions in the application.	✓	
Avoid overloading components with multiple input types.	✓	
Incorporate simple and quick signup process in the application.	✓	✓
Incorporate one feature on interface or one question with a progress bar in the application. This is sometimes referred to as gamification.	✓	
Verify the users of the application e.g., through an original selfie.	✓	✓
Avoid animations and marquees (e.g., text moving from top to bottom) in the application.	(✓)	(✓)
Incorporate fraud protection features in the application.	✓	✓

Recommendations in bold *are also presented as example design patterns in Section 4 in this chapter.*

Table 1.
Synthesised recommendations extracted from think aloud protocol and digital content extraction.

Raters	Author 1 and 2	Author 1 and 3
Main Category	**Cohen's Kappa (κ)**	**Cohen's Kappa (κ)**
Usability	0.698	0.809

Table 2.
Cohen's kappa score for mapping of recommendations.

develop smartphone applications for tech-savvy OAs. Some of these are described earlier along with sample quotations from the participants. There were a total of 131 recommendations for Usability with some overlapping [36]. However, for the sake of preciseness, a subset of synthesised versions of these recommendations mapped against the source from which they were derived are presented in **Table 1**. A total of 14 design patterns were created from these recommendations, and 3 of them are presented in this study.

Each unique recommendation was mapped by all authors against the 7 Usability categories outlined by Morville's Usability honeycomb and then validated using Inter-rater reliability as described in Section 2. The scores for inter-rater reliability exceeded 0.61, indicating a high level of reliability. The values of Cohen Kappa's co-efficient (κ) for Usability, and all raters (author 1, 2 and 3) are shown in **Table 2**.

4. Transform to design patterns

We developed numerous design patterns for Usability after conducting Think Aloud and Digital Content Extraction. Patterns, derived from these, also include supporting literature, thus providing further information about how to develop applications for OAs. Other authors have used patterns to describe usability design recommendations, such as [20, 26]. We build on their work by applying a range of recommendations in this format. Design patterns are shown to be beneficial [60]:

Section	Description
Problem	Cognitive decline is one of the common issue related to ageing, making it difficult for older adults to remember complex things. In the context of technology, passwords are troublesome for older adults [1].
Rationale	The decrease in memory causes older adults to forget and ultimately leads to demotivation. This, in turn, acts as a major factor to stop using the system [1].
Solution	1. Make the app accessible without the need for a password. Use alternative access mechanisms such as bio-metrics, finger prints or face recognition or text message to phone number [1].
Type	Usability
Sub-type	Usable
Related Patterns	Incorporate one feature on interface or one question with a progress bar in the application. This is sometimes referred to as gamification.
References	[1] Boyd, K., Bond, R.R., Nugent, C.D. and Donnelly, M.P., 2018. EasiSocial: Recommendations in the development and training of social media tools for older people. In EasiSocial: Recommendations in the development and training of social media tools for older people.

Table 3.
Make app more accessible with easy log-in id - e.g. fingerprint instead of password.

- Improve communication between stakeholders through providing a common vocabulary e.g., developers and maintainers.

Section	Description
Problem	Older adults get confused by the fast moving objects or content on the smartphone applications. This is mainly due to difficulty associated with the level of perception [1–2].
Rationale	Fast moving objects and animations make it challenging for older adults to concentrate. They feel themselves as less able to understand the complexities of the applications and eventually stop using it [1–2]
Solution	1.Avoid animations and marquees (e.g., text moving from top to bottom) in the application [1–2].
Type	Usability
Sub-type	Desirable
Related Patterns	Allow the older adults to choose a preferred theme in the application.
References	[1] De Barros, A.C., Leitão, R. and Ribeiro, J., 2014. Design and evaluation of a mobile user interface for older adults: navigation, interaction and visual design recommendations. Procedia ComputerScience, 27, pp.369–378. [2] Nurgalieva, L., Laconich, J.J.J., Baez, M., Casati, F. and Marchese, M., 2019. A systematic literature review of research-derived touchscreen design guidelines for older adults. IEEE Access, 7, pp.22035–22058.

Table 4.
Avoid animations and marquees (e.g. text moving from top to bottom) in the application.

Section	Description
Problem	Older adults suffering from auditory problems might not be able to hear important notifications or voice-enabled conversation by the application [1–4].
Rationale	Hearing loss is one of the key problem that is associated with ageing and the applications should be designed for individuals who suffer from it. This will make the application inclusive, accessible and easy to use [1–4]
Solution	1.Incorporate auto-captioning for disabled users in the app [1–4]
Type	Usability
Sub-type	Accessible
Related Patterns	Make the tap sensitivity high to cater for older adults with dry and husky fingers.
References	[1] De Barros, A.C., Leitão, R. and Ribeiro, J., 2014. Design and evaluation of a mobile user interface for older adults: navigation, interaction and visual design recommendations. Procedia ComputerScience, 27, pp.369–378. [2] Nurgalieva, L., Laconich, J.J.J., Baez, M., Casati, F. and Marchese, M., 2019. A systematic literature review of research-derived touchscreen design guidelines for older adults. IEEE Access, 7, pp.22035–22058. [3] Al-Razgan, M.S., Al-Khalifa, H.S., Al-Shahrani, M.D. and AlAjmi, H.H., 2012, November. Touch-based mobile phone interface guidelines and design recommendations for elderly people: A survey of the literature. In International Conference on Neural Information Processing (pp. 568–574). Springer, Berlin, Heidelberg. [4] Leitão, R. and Silva, P.A., 2012. Target and spacing sizes for smartphone user interfaces for older adults: design patterns based on an evaluation with users.

Table 5.
Incorporate auto captioning for disabled users in the app.

- Improve programmer productivity and program quality.
- Encourage best practices even among experienced designers.

Due to limitations on the size and volume of this chapter, we present a subset of three example design patterns covering sub categories of usability, i.e. Usable, Desirable and Accessible, in **Tables 3–5**. The remaining set of design patterns can be accessed at [36]. Thus, these design patterns are the answer to the research question posed earlier:

Research Question - What do tech-savvy older adults expect from smart-phone applications?

5. Threats to validity

It is imperative to consider limitations and threats to validity of research [61]. This is particularly important for empirical studies in software engineering, where there is often a multitude of possible threats [62]. First and foremost, the targeted population of this study was tech-savvy older adults, so the findings might not be applicable as-is to age-related counterparts who experience further difficulties in learning and adopting new technology. However, additional findings which included non-tech savvy OAs are presented in [36]. Moreover, for construct validity, field notes taken during the two studies were added to the data collection form on the same day of the sessions. To improve participants' understanding, an information sheet was provided written in English, particularly for Think Aloud sessions. Moreover, Digital Content Extraction involved older adults from across the globe, which may have enhanced the external validity of this research. A known problem with every form of research is that the experience of the researcher can influence both the data collection and analysis positively or negatively. We tried to reduce this bias by involving all three researchers in evaluating the codes and by checking the criteria for data selection. In the context of Think Aloud sessions, the setting was only an approximation of a real-life situation, not the actual day-to-day life usage of technology by older adults. The primary limitation of the examination of online ageing forums was the focus on two forums only. Also, a limited sample of applications was selected from the mainstream app stores using stratified sampling. Despite the limitations, the results follow from the data collected via empirical studies. For reliability, the evaluation of recommendations, through the use of inter-rater reliability, added confidence that the study would yield similar results if other researchers were to replicate our methods.

6. Conclusion

The empirical studies conducted with tech-savvy OAs highlighted their specific expectations from smartphone applications. Through conducting Think Aloud Sessions and Digital Content Extraction we elicited a set of requirements that we categorised according to Peter Morville's Usability Honeycomb [37]. To provide a level of objectivity to our classification, we validated our results in a series of inter-rater reliability tests. The output of this study is a set of recommendations for developing smartphone applications for a tech-savvy ageing population. Finally, we transformed the recommendations into 14 design patterns (three of which are presented in **Tables 3–5**). These patterns will help make these recommendations usable for practitioners or for pedagogy. For academia, the research results aim to provide an understanding of the recommendations for developing smartphone applications for OAs. For industry, the recommendations in the design patterns aim to help the development of usable smartphone applications for tech-savvy older adults, from development to testing of applications. We do not claim that the results presented are complete, but our research study provides significant results, and the design patterns in particular, are ready to be used by both developers and educators.

Acknowledgements

This work was supported, in part, by Science Foundation Ireland grant 13/RC/2094, co-funded under the European Regional Development Fund through the Southern & Eastern Regional Operational Programme to Lero - the Science Foundation Ireland Research Centre for Software (www.lero.ie) and IBM Ireland. We also thank the participants who gave their time in our 'think aloud' sessions.

Author details

Bilal Ahmad[1], Ita Richardson[1,2] and Sarah Beecham[1*]

1 Lero - The Irish Software Research Centre, University of Limerick, Ireland

2 Ageing Research Centre, Health Research Institute, Ireland

*Address all correspondence to: sarah.beecham@lero.ie

References

[1] Ageing Working Group of the Economic Policy Committee. 2018 Ageing Report: Policy challenges for ageing societies. European Commission Directorate-General for Economic and Financial Affairs; 2018. Available from: https://ec.europa.eu/info/news/economy-finance/policy-implications.

[2] Central Statistics Office Ireland. An Age Profile of Ireland; 2017. Web page (accessed: 26 March 2019). Available from: http://www.cso.ie/en/.

[3] European Commission. The 2018 Ageing Report Economic and Budgetary Projections for the 28 EU Member States (2016–2070). Economic and Financial Affairs; 2018. Available from: https://ec.europa.eu/info/sites.

[4] Fisk D, Rogers W, Charness N, Czaja S, Sharit J. Principles and creative human factors approaches. Boca Raton, Fla: Taylor and Francis Group LLC; 2009.

[5] Barrett A, Burke H, Cronin H, Hickey A, Kamiya Y, Kenny RA, et al.. Fifty plus in Ireland 2011: first results from the Irish Longitudinal Study on Ageing (TILDA); 2011.

[6] Siegel C, Dorner TE. Information technologies for active and assisted living—Influences to the quality of life of an ageing society. International journal of medical informatics. 2017; 100:32–45.

[7] Lee C, Coughlin JF. PERSPECTIVE: Older adults' adoption of technology: an integrated approach to identifying determinants and barriers. Journal of Product Innovation Management. 2015; 32(5):747–759.

[8] Berenguer A, Goncalves J, Hosio S, Ferreira D, Anagnostopoulos T, Kostakos V. Are Smartphones Ubiquitous?: An in-depth survey of smartphone adoption by seniors. IEEE Consumer Electronics Magazine. 2016;6 (1):104–110.

[9] Bull C, Harding M, Mikusz M, Knowles B, Davies N, Hayes N. Designing Mobile Applications for Older Adults. HotMobile. 2018.

[10] Wang JW. Retrieving critical design factor of ebook for older people in Taiwan. Telematics and Informatics. 2018;35(7):2016–2027.

[11] Knowles BH, Hanson VL. The Wisdom of Older Technology (Non-) Users. Communications of the ACM. 2018;61(3):72–77.

[12] Ahmad B, Richardson I, McLoughlin S, Beecham S. Assessing the level of adoption of a social network system for older adults. In: Proceedings of the 32nd International BCS Human Computer Interaction Conference 32; 2018. p. 1–5.

[13] Petrovčič A, Rogelj A, Dolničar V. Smart but not adapted enough: Heuristic evaluation of smartphone launchers with an adapted interface and assistive technologies for older adults. Computers in Human Behavior. 2018; 79:123–136.

[14] Dodd C, Athauda R, Adam M. Designing user interfaces for the elderly: a systematic literature review. In: Proceedings of the Australasian Conference on Information Systems; 2017. p. 1–11.

[15] Al-Razgan MS, Al-Khalifa HS, Al-Shahrani MD, AlAjmi HH. Touch-based mobile phone interface guidelines and design recommendations for elderly people: A survey of the literature. In: International Conference on Neural Information Processing. Springer; 2012. p. 568–574.

[16] Calak P. Smartphone evaluation heuristics for older adults [Ph.D. thesis]. University of Guelph; 2013.

[17] Silva PA, Holden K, Nii A. Smartphones, smart seniors, but not-so-smart apps: A heuristic evaluation of fitness apps. In: International Conference on Augmented Cognition. Springer; 2014. p. 347–358.

[18] Petrovčič A, Taipale S, Rogelj A, Dolničar V. Design of mobile phones for older adults: An empirical analysis of design guidelines and checklists for feature phones and smartphones. International Journal of Human–Computer Interaction. 2018;34(3):251–264.

[19] de Barros AC, Leitão R, Ribeiro J. Design and evaluation of a mobile user interface for older adults: navigation, interaction and visual design recommendations. Procedia Computer Science. 2014;27:369–378.

[20] Leitão R, Silva PA. Target and spacing sizes for smartphone user interfaces for older adults: design patterns based on an evaluation with users. In: Proceedings of the 19th Conference on Pattern Languages of Programs. The Hillside Group; 2012. p. 5.

[21] Díaz-Bossini JM, Moreno L. Accessibility to mobile interfaces for older people. Procedia Computer Science. 2014;27:57–66.

[22] Coelho J, Duarte C. A literature survey on older adults' use of social network services and social applications. Computers in Human Behavior. 2016; 58:187–205.

[23] Nurgalieva L, Laconich JJJ, Baez M, Casati F, Marchese M. A systematic literature review of research-derived touchscreen design guidelines for older adults. IEEE Access. 2019;7:22035–22058.

[24] Web Accessibility Initiative WAI. Accessibility, Usability, and Inclusion; 2016. Web page (accessed: 2 April 2019). Available from: https://www.w3.org/WAI/.

[25] Nurgalieva L, Laconich JJJ, Baez M, Casati F, Marchese M. Designing for older adults: review of touchscreen design guidelines. In: Workshop on Designing Mobile Interactions for the Ageing Populations; 2017. ArXiv preprint arXiv:1703.06317.

[26] Luna-Garcia H, Mendoza-Gonzalez R, Alvarez-Rodriguez FJ. Design patterns to enhance accessibility and use of social applications for older adults. Comunicar. 2015;23(45):85–94.

[27] Ahmad B, Richardson I, Beecham S. A systematic literature review of social network systems for older adults. In: International Conference on Product-Focused Software Process Improvement. Springer; 2017. p. 482–496.

[28] Lewis C. Using the 'thinking-aloud' method in cognitive interface design. IBM TJ Watson Research Center; 1982. RC9265.

[29] Dumas JS. User-based evaluations. In: The human-computer interaction handbook. L. Erlbaum Associates Inc.; 2002. p. 1093–1117.

[30] Lewis JR. Usability testing. In: Handbook of human factors and ergonomics. vol. 12. Wiley, New York; 2006. p. e30.

[31] Nielsen J. Thinking Aloud: The Number 1 Usability Tool; 2012. Web page (accessed: 23 September 2019). Available from: https://www.nngroup.com/articles/thinking-aloud-the-1-usability-tool/.

[32] den Haak MJV, de Jong MD, Schellens PJ. Employing think-aloud protocols and constructive interaction to

test the usability of online library catalogues: a methodological comparison. Interacting with computers. 2004;16(6):1153–1170.

[33] Clemmensen T, Hertzum M, Hornbæk K, Shi Q, Yammiyavar P. Cultural cognition in the thinking-aloud method for usability evaluation. In: Icis 2008 Proceedings; 2008. p. 189.

[34] Hewson C, Stewart DW. Internet research methods. In: Wiley StatsRef: Statistics reference online. Wiley Online Library; 2014. p. 1–6.

[35] Im EO, Chee W. Practical guidelines for qualitative research using online forums. Computers, informatics, nursing: CIN. 2012;30(11):604.

[36] Ahmad B. ReDEAP: Recommendations for developing smartphone applications for an ageing population [Ph.D. thesis]. University of Limerick; 2020. Available from: https://ulir.ul.ie/handle/10344/9283.

[37] Morville P. User Experience Design; 2013. Web page (accessed: 16 April 2019). Available from: http://semanticstudios.com/.

[38] Nielsen JM, Mack K. RL (1994) Usability Inspection Methods. New York: John Wiley & Sons; 1994.

[39] Quesenbery W, Design WI. Dimensions of usability: Defining the conversation, driving the process. In: UPA 2003 Conference; 2003. p. 23–27.

[40] Hartmann DP. Considerations in the choice of interobserver reliability estimates. Journal of applied behavior analysis. 1977;10(1):103–116.

[41] Suen HK, Lee PS. Effects of the use of percentage agreement on behavioral observation reliabilities: A reassessment. Journal of Psychopathology and Behavioral Assessment. 1985;7(3): 221–234.

[42] Bennett EM, Alpert R, Goldstein A. Communications through limited-response questioning. Public Opinion Quarterly. 1954;18(3):303–308.

[43] Zwick R. Another look at interrater agreement. Psychological Bulletin. 1988; 103(3):374.

[44] Cicchetti DV, Shoinralter D, Tyrer PJ. The effect of number of rating scale categories on levels of interrater reliability: A Monte Carlo investigation. Applied Psychological Measurement. 1985;9(1):31–36.

[45] Umesh UN, Peterson RA, Sauber MH. Interjudge agreement and the maximum value of kappa. Educational and Psychological Measurement. 1989;49(4):835–850.

[46] Gordis L. Epidemiology. WB Saunders Company, Pennsylvania; 1996.

[47] Cohen J. Weighted kappa: Nominal scale agreement provision for scaled disagreement or partial credit. Psychological bulletin. 1968;70(4):213.

[48] McHugh ML. Interrater reliability: the kappa statistic. Biochemia medica: Biochemia medica. 2012;22(3):276–282.

[49] Zhang C, Budgen D. What do we know about the effectiveness of software design patterns? IEEE Transactions on Software Engineering. 2011;38(5):1213–1231.

[50] Bieman JM, Straw G, Wang H, Munger PW, Alexander RT. Design patterns and change proneness: An examination of five evolving systems. In: Proceedings. 5th International Workshop on Enterprise Networking and Computing in Healthcare Industry (IEEE Cat. No. 03EX717). IEEE; 2004. p. 40–49.

[51] Alexander C. A pattern language: towns, buildings, construction. Oxford university press; 1977.

[52] Gamma E. Design patterns: elements of reusable object-oriented software. Pearson Education India; 1995.

[53] Noll J, Richardson I, Beecham S. Patternizing GSD research: Maintainable decision support for global software development. In: 2014 IEEE 9th International Conference on Global Software Engineering. IEEE; 2014. p. 110–115.

[54] Ambler S. The process patterns resource page; 2000. Web page (accessed: 22 December 2019). Available from: http://www.ambysoft. com/processPatternsPage.html.

[55] Cloutier R. Applicability of patterns to architecting complex systems [Ph.D. thesis]. Stevens Institute of Technology Hoboken, NJ USA; 2006.

[56] Dickie C, Forde P. The hidden potential of organisational planning: matching Alexander's theory of patterns to practice. Connecting theory and practice. 2007:1–28.

[57] Creswell JW, Creswell JD. Research design: Qualitative, quantitative, and mixed methods approaches. Sage publications; 2017.

[58] Braun V, Clarke V. Using thematic analysis in psychology. Qualitative research in psychology. 2006;3(2):77–101.

[59] Ahmad B, Richardson I, McLoughlin S, Beecham S. Older adults' interaction with mobile devices in ireland: A survey. In: Mobile Privacy and Security for an Ageing Population: Workshop at MobileHCI 2018: September 3 Barcelona, Spain; 2018. Available from: https://csalsa.gitlab.io/mobilehciageing/submissions.htm.

[60] Prechelt L, Unger-Lamprecht B, Philippsen M, Tichy WF. Two controlled experiments assessing the usefulness of design pattern documentation in program maintenance. IEEE Transactions on Software Engineering. 2002;28(6):595–606.

[61] Easterbrook S, Singer J, Storey MA, Damian D. Selecting empirical methods for software engineering research. In: Guide to advanced empirical software engineering. Springer; 2008. p. 285–311.

[62] Feldt R, Magazinius A. Validity threats in empirical software engineering research-an initial survey. In: Proceedings of the International Conference on Software Engineering and Knowledge Engineering; 2010. p. 374–379.

Chapter 5

Indian Language Compatible Intelligent User Interfaces

Sanghamitra Mohanty

Abstract

Implicit Human Computer Interaction to explicit context-dependent knowledge processing to meet the requirements of intelligent human being in the present day situation has provoked scientist towards the development of Intelligent User Interfaces to face the society with a demand of smart devices, embedded to the smart environment and dynamic activities. Context-aware situation deals with location and situation focused context like virtual car navigation to context with respect to traffic to climate to the human connected with it immediately and afterwards.

Keywords: HCI, IUI, AI, Russel's Model, Plutchik's Model, Table Top Devices, command based TTS, Speech enabled OCR

1. Introduction

The present day busy and techno savvy life of the people and more with youngsters are provoking people to think of utilizing the brain power more than the other body parts as used in the so to say busy and intelligent life style. Technology is being used by all much more than ever before. It has already been tried with Human Computer Interaction (HCI) applications successfully [1]. The present day presentation, reading, writing and speaking are expected to be done in smart system, where multiple works can be done simultaneously using finger tips, gesture, posture, vision and speech or any other system, which is user friendly and controlled by the brain signals a bit directly giving emphasis to cognition management before execution. Or we can say that in IUI the net impact is the influence of Artificial Intelligence (AI) on Human Computer Interaction on the activities of HCI. Instead of the traditional interfaces one can use Intelligent Interfaces to interact with the day to day life activities starting from study table to any type of action (**Figure 1**).

It is assumed and accepted that Human Computer Interaction has been observed as implicit interaction with respect to traditional communication. Few years back we are habituated to use traditional communication system like telephone, letter writing, personal meeting etc. Later with the development of computers we are in a situation of using the computer, mailing service and mobile phones etc. But as we are entering more towards automation the communication system that we are going to use in near future is totally advanced with respect to the Human Computer Interaction system (HCI) [2–4]. Automation along with the Artificial Intelligence has influenced our life so much that we are bound to move with the fast moving system in the world (**Figure 2**).

It is observed that the availability of automation and speedy life style has forced us to opt for automation. Presently every country is having supportive infrastructure to manage this type of automation. Different type of Interfaces is the basic

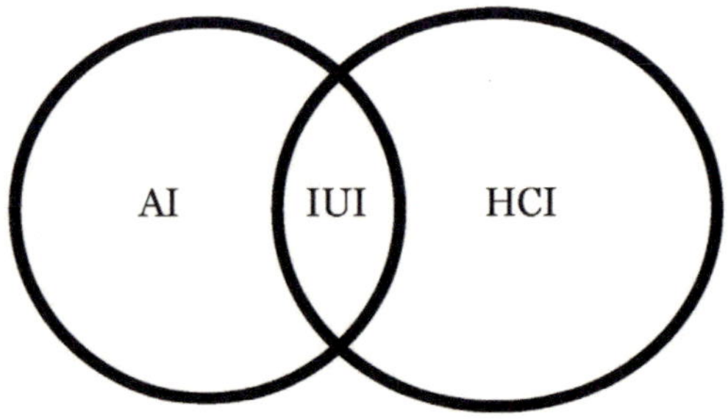

Figure 1.
Interdependency of interactions with techniques, machines and interfaces [1].

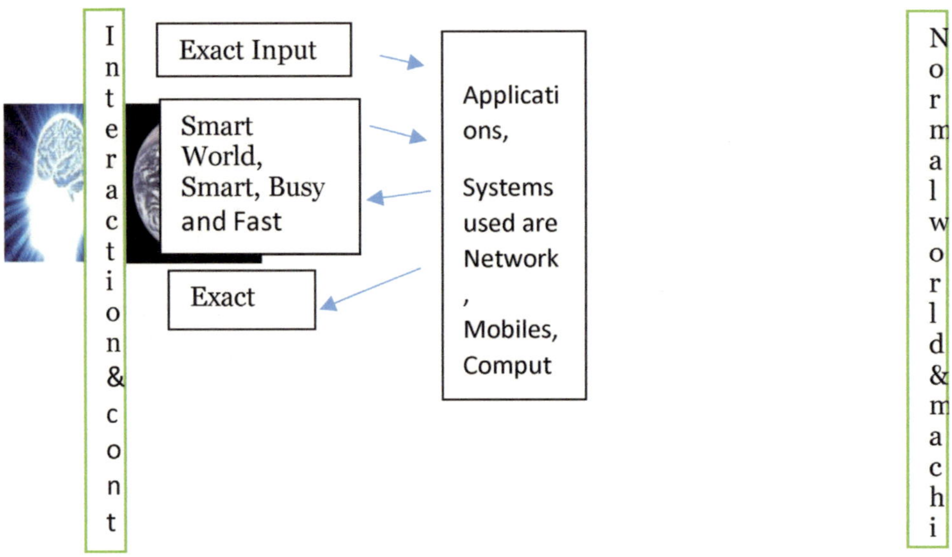

Figure 2.
The concept diagram for the need and basics of intelligent user interface [2–4].

need to meet the challenges of the smart world through which we are passing through presently. Among the smart Intelligent Interfaces different type of intelligent interfaces may include devices like signal based, gesture based, and speech based but not limited to these types. In a specific way it may be mentioned that IUI uses the following platforms to be effective and useful [5] (**Figure 3**).

The challenges to this device development and utilization are many fold. Among those some very important issues are

1. Illiteracy
2. English Illiteracy
3. Unavailability of Network
4. Unavailability of Suitable Devices
5. Language compatibility

and many more human centric issues, which need attention to provide solutions.

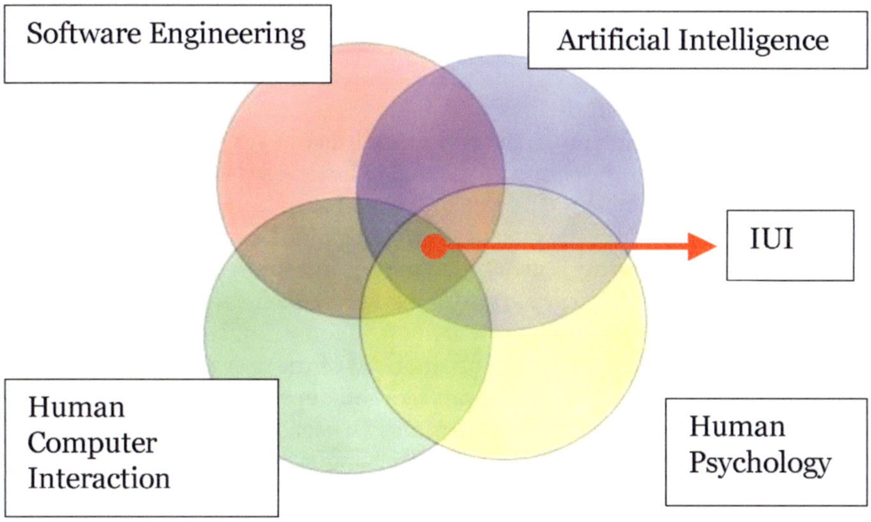

Figure 3.
Components of intelligent user interfaces [5].

Intelligent user interfaces deal with interactive human like interaction to solve any problem. Most effectively without typing can a person deal with the computer and its usage is the challenge of the day.

2. Need of intelligent user interfaces

Explosive materials availability and use in different sectors like national, international and global information extraction and exchange platform. Available present facilities or devices are not enough for futuristic use.

To provide solutions to such situations development of Intelligent User Interfaces are a must. Such Interfaces may include technological concepts like

i. Image Processing (Gesture)

ii. Speech Processing (Command mode)

iii. Sensors like light, sound, temperature, tactile and pressure sensors to make the Intelligent Interfaces interactive with gesture and posture without writing code or operating with human interventions (tactile sensor for touch sensitive interactions etc.)

iv. Internet of Things (IOT) compatible devices etc.

These factors have motivated the users of different levels to utilize the Multimedia facilities towards the interface development to the challenges due to the complexities in connectivity like telephone network, wireless and satellite communications. Multimedia solutions are a part of the solutions to such problems. Such efforts are supported by the improved microprocessor and storage facilities available now. Such device development may be included in all applications through different WIMP (windows, icons, menus, and pointing) interfaces [6]. It was realized that the real world problem can be better solved by providing Intelligent Interfaces.

Lot of research activities are being carried out and are also presently getting improved due to the dynamic growth of requirements both by human being and machine providing services to mankind in the field of Human Computer Interaction. Techniques embedded with Artificial Intelligence and Machine Learning in the physical devices leads towards the design and development of Intelligent User Interfaces. The Interfaces should be adaptive and adaptable to be intelligent [6, 7].

With respect to implicit Human Computer Interface development in anticipation to the behavior of the user the context expected to be developed the interfaces are to be designed. Most importantly the sensor output based interactions and context dependency can be solved for the purpose using Machine Learning Techniques as well as rule based systems. Thus the sensor based output interpretation and its generic user based modelling to use at large and in particular are challenges for the Intelligent User Interface users as their requirement is dynamic and sources of knowledge in streaming a type of Big Data. It also takes care of emotions of the user and acts accordingly. Context based factors are user based factors like and depends on Physical Environment like location, infrastructure and Human factors like user, social environment. Some of the Interfaces developed by the author in Indian Languages may be discussed here.

Any situation which deals with user and its action and provides output in the form of action which is proactive and functions with prior and appropriate anticipation of the result towards the user need is like motion detector in door opener and locker, beginning of an escalator or conveyor belt, smart communication devices like smart mobiles or tablets which changes it operational conveniences while moving in railway stations, shopping malls long corridors which enables user to pass through being unmindful towards the path. Making on and off of the rooms and room air-conditioners and adjusting the temperature inside the room may be a conference room even is the required setups which needs intelligence interfaces.

It is quite important to find the relationships between the intelligence in a system and the user interaction, which is called the intelligent processing found in the user interface(s) of the system. Such a system should meet the conditions namely effective, natural, or otherwise appropriate interaction of users with the system [8]. Here mainly the system helps the user with pseudo human communication system like speech, gesture or any other form of communication like human beings. Here the interfaces function like an intelligent human being to assist the user to function as per his/her will. This supports the interaction system like human to human interaction with the specially developed system with intelligent interfaces.

Further to say that implicit human computer interaction is an important aspect towards the development of Intelligent Interfaces. The most complex factor which is cognition oriented is computing the human behavior like recognition, interpretation, modelling processing and synthesis of human affects the emotions. Recognized human computer interaction can help in developing effective Intelligent Interfaces. Emotion is of great importance while developing Intelligent User Interfaces as use of those purely depends on the emotions of the users.

Different types of emotions are described by different scientists in due course. Some of those are mentioned below for reference.

Under discrete model Ekman's Model is well established in treating emotions in terms of discrete and different construct for the analysis of emotion. As per Paul Ekman Model there are seven basic emotions namely Happiness, Sadness, Fear, Disgust, Anger, Contempt and Surprise.

Under Dimensional Model emotions are described mostly through Russel's Model and Plutchik's Model.

i. Russel's Model by James A Russell describes in two dimensional circular space like arousal and valence with as much as twenty-eight types of emotion having five major classes namely Happy, Relaxed, Sad, Angry and Normal.

ii. Plutchik's Model by Robert Plutchik [9] has eight standard models described as an extension of Russel's Model where the eight emotions are anger, anticipation, joy, trust, fear, surprise, sadness and disgust (**Figures 4** and 5).

Pultchik's Wheel of Emotion is described as the wheel of emotion classification and have eight types of resultant emotions. To identify the exact emotion of the user to work accordingly is a sensitive issue.

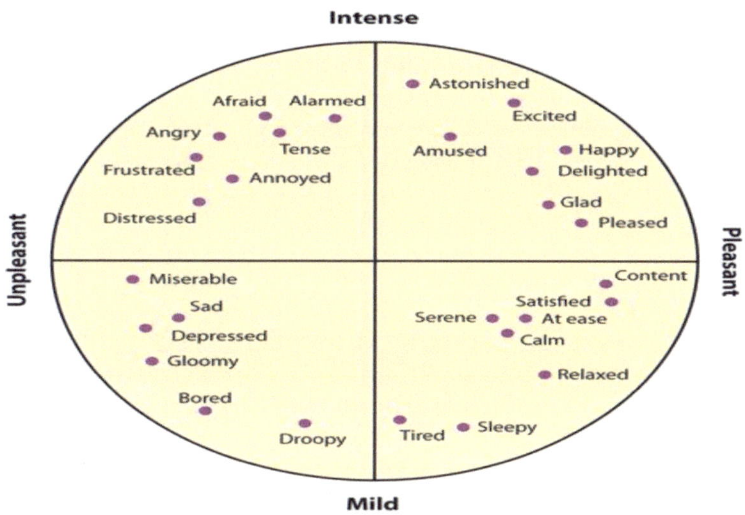

Figure 4.
Russel's model of emotions, where 28 types of emotions are displayed [10].

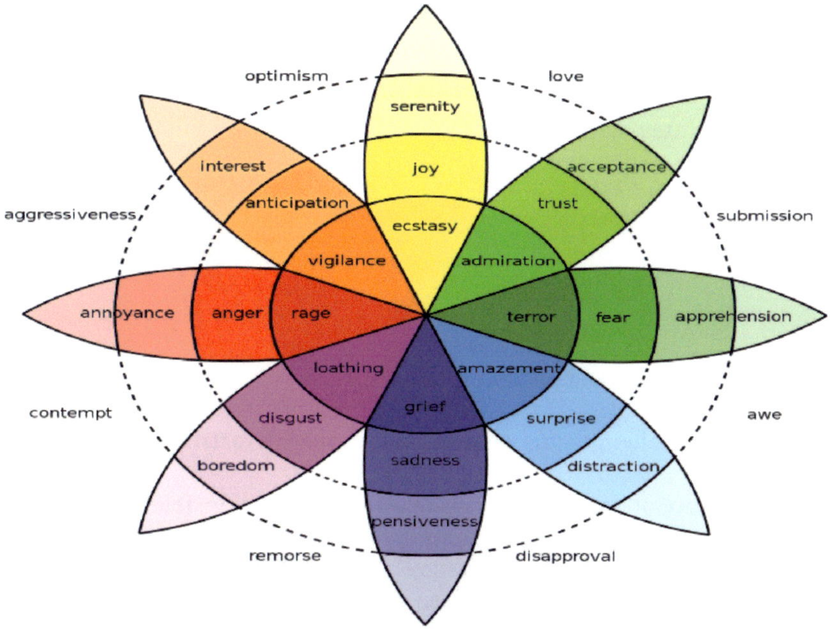

Figure 5.
Pultchik's wheel of emotion showing major eight types of emotion observed [9].

It is quite important to find the relationships between the intelligence in a system and the user interaction, which is called the intelligent processing is found in the user interface(s) of the system. Such a system should meet the conditions namely effective, natural, or otherwise appropriate interaction of users with the system. Sine here mainly the system helps with human like communication system like speech, gesture or any other form of communication like human beings. Here the interfaces function like an intelligent human being to assist the user to function as per his/her will. This supports the interaction system like human to human interaction with the specially developed system with intelligent interfaces.

Intelligent user Interface systems should have mainly the following criteria like: recommender systems, products, documents, or any other consequential follow-up action. This involves intelligent technology to support information retrieval, appropriate learning environments with learning assistance interface agent with capacity to monitor a situation intelligently assessing the situation and acts for the user of the system using the knowledgebase created with the domain. The main factor of Intelligent Interfaces is the cognition and the agent has to act and react as per the emotions embedded into the need of the user.

3. Challenges in developing intelligent user Interface

The Intelligent Agent that functions behind the IUI has at least the following characteristics [5].

a. Input analysis

b. Output functioning

c. Agent based interaction

d. Multimodal output presentation interfaces

e. Well defined model based interfaces

f. Image based Interaction Interface

g. Speech based interaction interface

h. Text analyzing capacity

i. Heuristic Searching capacity

j. Intelligent output layout designing capacity

k. Handling the challenges due to multilingual speech instructions

The performance of an intelligent component may be sensible to failure leading to inappropriate interpretations of commands leading to action and appropriate reaction as Intelligent processing is often relatively difficult for users to predict, understand, and control a demanding situation online. If the demand is bit more elongated depending on relatively extensive information about the users, the intelligent systems may raise certain types of privacy and security risks as private data may not be easy due to security point.

Major challenge to such a situation is evaluating the order of performance of the algorithms used in design and execution of the Intelligent Interfaces. This involves Smart speech processing for talking to the smart human user, Image processing for viewing the activities of the smart human user and Natural Language Processing for understanding the language of a human who is not aware of the language the system is dealing with. As such system will be immediate use of the elderly people and small children. Besides these the factor that influences more is the environment which, controls the functioning of the Intelligent User Interfaces mostly as it changes its features like facilities with respect to user interaction accordingly.

To have Smart Eyes, Ears and Skin (touch) for the IUI are the foremost devices to be developed to enable it to act as the IUI in true sense. Besides the sensor based applications are a basic need for such a system. Presently sensors are operational through Internet. Most of the activities can be controlled by such devices called Internet of Things or IOT. These are the smallest unit in IUI for the largest objective satisfaction. The most important point here to use such devices in IUI is the Security in using IOT as it is internet controlled Cyber intervention are very much expected [11].

4. Different models for the intelligent user interface

User Interfaces are developed using mostly languages like XML, UML and UIML etc.as it deals with different types of users and their requirements in open mode [11].

There are models which are used towards the development of these Intelligent Interfaces.

a. First one is the Domain Model, which deals with data storage aspects as per the need of the users to complete an assigned work.

b. Next is the Task Model which is designed to represent the tasks the users may be allowed to perform taking into consideration the time factor of execution. It is mostly ConcurTaskTRees notation based model.

c. Vanderdonckt and Bodart described the Abstract User Interface Model which includes the user interface expressed in terms of abstract interaction objects [6]. This model is platform independent like Desktop, Laptop, PDA and Smart Phones and modality independent like voice or image input and output.

d. Then is the Concrete User Interface Model which represents the user interface, which is a composition concrete interaction objects [6]. Here the final user interface is developed to interact with the specified users of the system to support the Human Computer Interaction.

e. Finally, the Context Model deals with the whole of the context those are to be stored towards the operation and management of the system. It is a combination of the user model, a platform model and an environment model.

f. This is a combination of Artificial Algorithms to Human Centered Computing or in short Human Centered Artificial Intelligence or HCAI.

5. Challenges for intelligent user interfaces

Success of IUI depends on many factors like Misinterpretation of Cognitive Interaction leading to incorrect understanding by user, erroneous interpretation

Figure 6.
Samsung SUR40 tabletop intelligent user Interface [15].

of instructions by the devices leading to controlling of the system by the user. Depending upon the length of the query demanded by the user the issue of security and privacy will be compromised leading to complex situation [12, 13].

The important factor leading the functioning of the IUI is the dynamism of the user demand depending upon the environment to which the system is exposed to at that time. The solution may be strong algorithm development according to the dynamism of the situation around the user.

One of the most used IUI will be the Tabletop Devices as Intelligent User Interface [14, 15]:

The Tabletop design is an example of IUI to be used in different fields like Education system and Ambient Assistance system as an Augmented Reality facility to cater the needs of the present day situation, when the students aa well as the elderly people are exposed to distance mode of learning and functioning with multimodal facilities around them. This will enhance the growth of interactive tabletop device development activities to cater to the need of academicians and industry people. This has happened due to the increase of interactive systems with digital support brining digital power to table surfaces to support human activities intelligently. To mention the core aspects of interactive tabletop computing devices it may be mentioned that (i) Hardware development of horizontal touch surfaces, (ii) Multimodal Interfacing facilities as per the need like touch, speech, voice and gesture facilities meeting the novel touch and modes of use by the user. A snap shot of the Samsung SUR40 Intelligent User Interface is given below [15] (**Figure 6**).

6. Interface design in Indian language perspectives

One can think of designing an interface intelligent enough to cater the needs of the user. But the category of the users is a big challenge towards the development of the devices. Intelligent User Interface design will mostly depend on the features of Speech Processing, Image Processing and Natural Language Processing Tools.

A screen shot of a Speech Processing Tool developed for Odia Language is given below. This based on the concatenative method of phones used in Odia Language. It helps a user to listen to any text written in Odia language being converted to speech form. It uses the Artificial Intelligence based techniques to cater to the need of the user. This technique has already been tested for many Indian Languages including Indian English. Here the phones, which is the smallest sound unit used in any language are identified and depending upon the pronunciation rules those are concatenated to form a word, thus a sentence. A text may be directly written or read from a file already available like office orders. It reads it out for the user [16–19] (**Figure 7**).

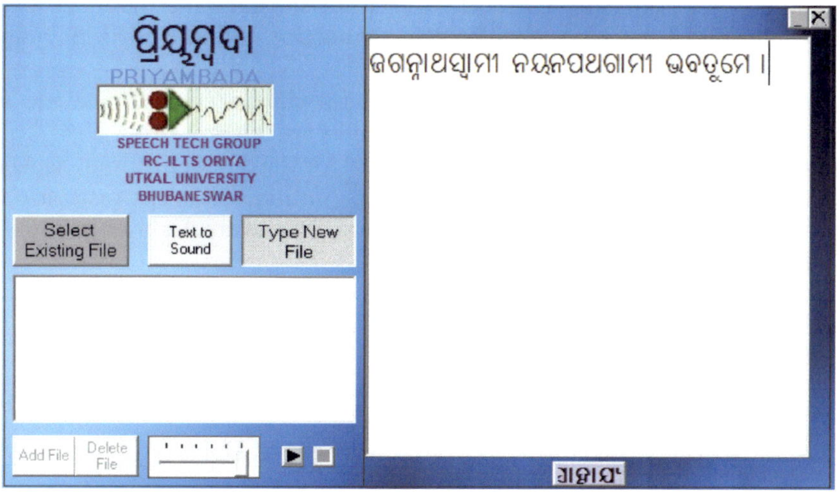

Figure 7.
Odia language phone based text-to-speech system [16].

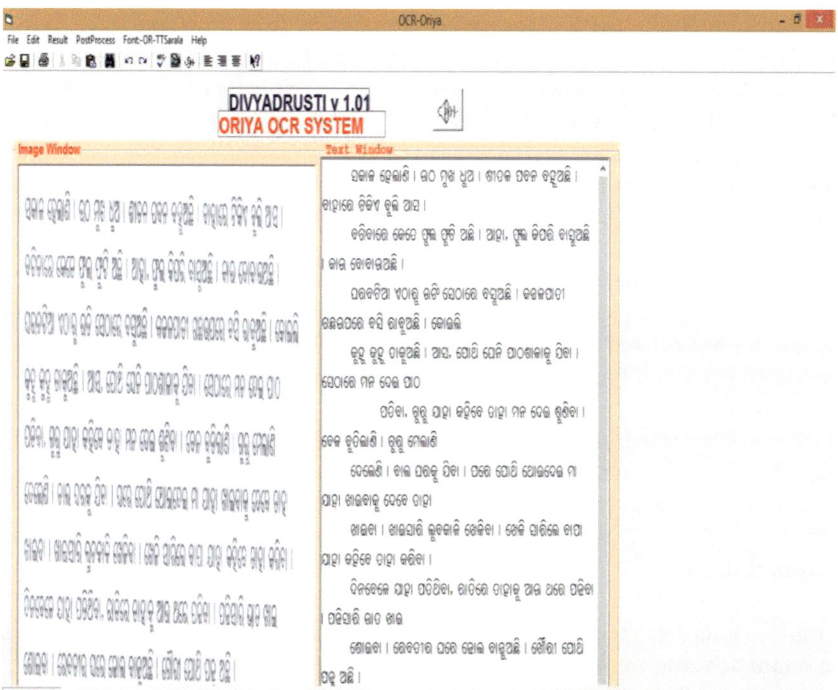

Figure 8.
Odia language optical character recognition system which performs on speech commands and reads out the output the OCRed text to the listen [29].

It is observed that many documents are in scanned document form which is not possible to link it to any Text-To-Speech System. Here is a successful attempt to read the content for the needy persons like illiterate and visually challenged [20–25].

A snap shot of the intelligent user interface for physically challenged is shown below, which works on mouse click as well as speech commands [26–30] (**Figure 8**).

For all such type of Intelligent Interface Development, Natural Language matters as the user may not be English Language Always. The problem to make it more effective towards global use the language factor consisting of Image and Speech are to be handled where the natural process of writing, speaking matters. Thus the factors to be considered are to be considered in connection with the Natural Language Processing aspect.

Here the author has used several user interfaces which are able to manage intelligently the commands or uses expressed in terms of speech or written form. To solve it and exactly interpret to convert it to command for the device different aspects of NLP like Machine Translation, Name Entity Recognition, Grammar restriction free command etc. free like natural speech in command mode are dealt with. Some of the works are mentioned in the papers [31–34].

Even the gesture is most important as any user will prefer to use gesture to express the mind as at cognition level the brain synthesizes the command and expresses it different form like Speech or Text.

7. Conclusion and immediate future challenges of IUI

While entering to Industrial Revolution 4.0, which is a combination of Automation and Internet, device development is a challenge. More of Automated devices are to be developed, which will use Sensor based commands and will execute it as per the need of the users. Thus IOT based device development is inevitable.

Such work involving the Sensor based includes but not limited to Multiple Sensor based Table Tops. It may be multimodal, multiuser and affordance of the table for multimodal operations accommodating the objects on the Tabletop. Here not only the Tabletops support hand and gesture mode operation but also tangible interaction and often needed visualization [35, 36].

Table Top applications have a massive application in viewing dominated working platform like photo or videos, brain teasing activities like game or competition besides normal computational work to advanced visualization of Dataset dynamically.

The important points for research to be handled for such devices are

a. Computer Supported Cooperative Work (CSCW), which is the computer-mediated group collaboration and social interactions

b. Human Computer Interaction (HCI) beyond the desktop setup to individual user actions and reactions.

c. Ubiquitous Computing (UbiComp) to accommodate variety of input data and output models.

d. Tangible User Interfaces (TEI) which involves interaction with physical objects on a tabletop and to ovoid those dynamically.

Considering the simple Table top application as one of the simple examples of IUI we observe that the following few parameters are to be taken into consideration while developing the devices. To achieve it mostly we have to go for IOT based services.

The other point to be handled by the scientists are

a. The Touch Panel, Sensor grid (electrodes) covered by front layer, which enables accurate detection of multi-touch, high positional accuracy.

b. The other point is Surface/projected capacitive solutions only work with finger or special stylus and not suited for large panels with slower transmission of electrical current. These items are mostly used in Smart Mobiles, which is under metamorphism as per the requirements and demands of the users.

c. Low-cost manufacturing, Low power consumption compared to other approaches.

d. Inputs requires pressure on the outer layer.

e. Reactive to the stylus or when user wearing gloves.

f. Reduced display quality due to the additional complex layers for execute the functioning details.

g. Compatible to Mobile Phones, PDAs, Digital Cameras or any other devices under development and use plan.

These are the future challenges which need quick solutions to meet the busy, preoccupied and smart users for their needs. Hope the trend of current research leading to such requirements can provide solutions to these problems as Intelligent User Interfaces are unavoidable in the future generation technology users. It is as dynamic as the need of the end user and the development is a challenge to the developers.

These are leading to the aspects of Augmented Reality and Virtual Reality, which is the Next Generation Computing.

Author details

Sanghamitra Mohanty
Department of Computer Science and Application, Utkal University, Bhubaneswar, India

*Address all correspondence to: sanghamitramohanty1@gmail.com

References

[1] "Intelligent User Interfaces: An Introduction", Morgan Kaufmann, RUIU, San Francisco, 1998, pp. 1-13

[2] "Next Generation Intelligent Interfaces: Implicit Human-Computer Interaction", Beat Signer, Vrije Universiteit Brussel.

[3] "Cross-Media Information Spaces and Architectures", Beat Signer, Vrije Universiteit Brussel.

[4] "Intelligent User Interfaces", Daniel Sonntag, German Research Centre for Artificial Intelligence (DFKI).

[5] "Intelligent User Interfaces: Past, Present and Future", V. López Jaquero, et.al., https://www.researchgate.net/publication/226527078

[6] "A Data Processing System for Air Defense", R. Everett, C. A. Zraket, and H. Bennington, SAGE: , 5(4), 1983.

[7] "Intelligent user interfaces: an introduction", Maybury M. In Proceedings of the 4th international Conference on intelligent User interfaces, 'IUI' 99. ACM Press, New York, 1999.

[8] "Model-Based Design and Evaluation of Interactive Applications", F. Paternò, Springer. 1999.

[9] "Summarizing Emotions from Text Using Plutchik's Wheel of Emotions", Advances in Intelligent Systems Research, volume 166, 7th.M. M. Abasi and Anatoly Beltiukov, Scientific Conference on Information Technologies for Intelligent Decision Making Support (ITIDS 2019).

[10] *"A Circumplex Model of Affect* J.A. Russel,, Journal of Personality and Social Psychology, 39(6), 1980, https://content.apa.org/doi/10.1037/h0077714

[11] "Encapsulating Knowledge for Intelligent Automatic Interaction Objects Selection", Vanderdonckt, J., Bodart, F. In ACM Proc. of the Conf. On Human Factors in Computing Systems INTERCHI'93 (Amsterdam, 24-29 April 1993), S. Ashlund, K. Mullet, A. Henderson, E. Hollnagel & T. White (Eds.), ACM Press, New York, 1993.

[12] "Steps to take before Intelligent User Interfaces become real", Kristina Hook, http://www.sics.se/~kia/, 2000.

[13] "Comprehensive Understanding of Intelligent User Interfaces", S. Shaikh, N. A. Khan and F. B. Solangi, (IJACSA) International Journal of Advanced Computer Science and Applications, Vol. 8, No. 6, 2017.

[14] "Next Generation User Interfaces Interactive Tabletops and Surfaces", Beat Signer, Vrije Universiteit Brussel

[15] "Interactive Tabletops and Surfaces - Lecture 07 – Next Generation User Interfaces (4018166FNR)", 2017.

[16] "Syllable based Indian language text to speech system", S. Mohanty, International Journal of Advances in Engineering & Technology, Vol. 1(2), 2011.

[17] "Use of SVM Classifier & MFCC in Speech Emotion Recognition System", Khyamamayee Dash and Prof. Sanghamitra Mohanty Bhoomika Panda, Debananda Padhi, IJARCSSE, 2012.

[18] "Speaker identification using mel frequency cepstralcoefficient and bpnn", Kshamamayee Dash, Debananda Padhi, Bhoomika Panda, Sanghamitra Mohanty, 2012.

[19] "Phonotactic model for spoken language identification in Indian language perspective", S. Mohanty, International Journal of Computer Applications, 2011.

[20] "Speaker Identification using SVM during Oriya Speech Recognition.",

Sanghamitra Mohanty, Basanta Kumar Swain, International Journal of Image, Graphics & Signal Processing, 2015.

[21] "Language identification using support vector machine", Sanghamitra Mohanty, Basanta Kumar Swain, OCOCOSDA, 2010.

[22] "Emotion recognition using fuzzy K-means from Oriya speech", Sanghamitra Mohanty, Basanta Kumar Swain, 2010 for International Conference [ACCTA-2010].

[23] "Language Identification-A Supportive Tool for Multilingual ASR in Indian Perspective", Sanghamitra Mohanty and Basanta Kumar Swain, ICACIE2019, 2019.

[24] "Isolated Word based Spoken Dialogue System using Odia Phones", Sanghamitra Mohanty and Basanta Kumar Swain, ICICT2019, 2019.

[25] "Isolated Word-based Spoken Dialogue System using Odia Phones", Sanghamitra Mohanty, Basanta Kumar Swain and Chiranji Lal Chowdhary, Recent Advances in Computer Science and Communications, Volume 14, Issue 6, 2021.

[26] "Pattern recognition in alphabets of Oriya language using Kohonen neural network", S. Mohanty, International journal of pattern recognition and artificial intelligence, World Scientific Publication 1997.

[27] "An efficient bilingual optical character recognition (English-Oriya) system for printed documents", S. Mohanty, H. N. Dasbebartta and T.K. Behera, Seventh International Conference on Advances in Pattern Recognition, IEEE Publications, 2009.

[28] "Oriya Character recognition using neural networks", Soumya Mishra, Debashish Nanda, Sanghamitra Mohanty, Special issue of IJCCT, 2010.

[29] "A complete OCR development system for Oriya script", S. Mohanty and H. K. Behera, SIMPLE, 2004.

[30] "Algorithm for segmenting script-dependent portion in a bilingual Optical Character Recognition system", Sanghamitra Mohanty and Himadri Nandini Das Bebartta, Pattern Recognition and Image Analysis, 27, 2017.

[31] "Semantic based text classification using wordnets: Indian language perspective", S Mohanty, PK Santi, Ranjeeta Mishra, RN Mohapatra, Sabyasachi Swain, Proceedings of the 3th International Global WordNejuh t Conf., South Jeju Island, Korea, 2006.

[32] "A two stage language independent named entity recognition for indian languages", S Biswas, MK Mishra, S Acharya Sitanath_biswas, S Mohanty, International Journal of Computer Science and Information Technologies, 2010.

[33] "A hybrid Oriya Named Entity Recognition System: harnessing the power of rule", Sitanath Biswas, SP Mishra, S Acharya, S Mohanty, International Journal of Artificial Intelligence and Expert Systems (IJAE), 2010.

[34] "Oriya Language Text Mining Using C5. 0 Algorithm", Sohag Sundar Nanda, Soumya Mishra, Sanghamitra Mohanty, (IJCSIT) International Journal of Computer Science and Information Technologies, Vol. 2 (1) , 2014.

[35] "Architecting next-generation user interfaces", Daniel Wigdor, Proceedings of the International Conference on Advanced Visual Interfaces, AVI 2010, Roma, Italy, 2010

[36] "What's Hot in Intelligent User Interface", Proceedings of the Thirtieth AAAI Conference on Artificial Intelligence (AAAI-16)", Shimei Pan, Oliver Brdiczka, Giuseppe Carenini and Duen Horng Chau, 2016.

Section 3

Development Methodologies and Techniques

Chapter 6

Negative UX-Based Approach for Deriving Sustainability Requirements

Nelly Condori-Fernandez, Marcela Quispe-Cruz, Alejandro Catala, Joao Araujo and Patricia Lago

Abstract

In this chapter, a Negative User Experience (NUX)-based method for deriving sustainability requirements of persuasive software systems is proposed. The method relies on the analysis of NUX assessment, and the exploitation of relationships between the SQ model and the PSD model, which are well-known models for sustainability-quality in software systems and persuasive system design respectively. To illustrate the method, a user study has been conducted involving people in their real working environments while using specific software intended to change their behavior for preventing or reducing repetitive strain injury (RSI). The method allowed us to discover thirteen requirements that contribute to social, technical and economic sustainability dimensions.

Keywords: usability, sustainability-quality model, quality attributes, PSD model, persuasive software systems, UX assessment

1. Introduction

Persuasive technology (PT) can be defined as "design, research, and analysis of interactive computing products created to change people's attitudes or behaviors" [1]. As technology can be used as a promoter of sustainable behavior, many studies have investigated the possibilities to persuade people within the context of environmental sustainability (e.g., increase consumers' awareness of energy consumption [2]. However, most of these studies have shortcomings that limit their long-term effectiveness. Although behavioral models (e.g., Transtheoretical Model of behavior change [3], the Goal-setting Theory [4], the Fogg Behavior Model [5]) are very useful for conceptualizing the impact of persuasive technology, most of them cannot easily be applied to the design or assessment of persuasive systems directly because they do not provide appropriate methodological support [6, 7]. For example, through a user experience assessment of existing persuasive software applications, Condori-Fernandez et al. [7] found that some relevant non-functional requirements had not been addressed, and consequently users experienced negatively in using such kinds of systems.

As the identification and management of non-functional requirements in software projects are challenging [8], various assessment models have been proposed for software product quality (e.g., ISO/IEC 25010 quality model). In the software

IntechOpen

engineering community, a software sustainability model consists of both sustainability-related requirements and quality requirements (e.g., [9–11]). Lago et al. [12] defined software sustainability based on a four-dimensional model that adds the technical dimension to the social, environmental and economic dimensions that already appear in the Brundtland report [13]. Condori-Fernandez and Lago [9] proposed a Sustainability-Quality (SQ) model for supporting the identification of quality requirements that contribute to the four-dimensional model of software-intensive systems[1]. The multidimensional approach of Becker et al. [15] adds the individual dimension to the four sustainability dimensions [12]. However, Calero et al. [10] define sustainability only in terms of energy consumption, resource optimization and perdurability, and they do not consider the individual, social, and economic dimensions.

According to Assefa and Frostell [16], for a system to be deemed socially sustainable, it should at minimum enjoy wider social acceptance. In this respect, ensuring the quality of User Experience (UX) is important for increasing the likelihood of accepting socially software systems (e.g., [17]).

In order to provide support for discovering non-functional requirements (NFR) that contribute to sustainability dimensions, we present a Negative User Experience (NUX)-based approach for deriving sustainability-quality requirements, with special emphasis on the social and technical dimensions. Then, the approach is applied in existing software applications designed for preventing RSI.

The following sections provide a detailed account of our work. Section 2 describes the SQ model and PSD model on which the NFR discovery approach is based. Section 3 presents the NUX based approach for deriving sustainability requirements. As a result of applying the approach, we present first the design of a user study. And the discovered nonfunctional requirements and features are reported in Section 5. Finally, we draw the conclusions.

2. Background

In this work, we adopted the PSD model [18] as the theoretical framework for our research, and the SQ model [9].

2.1 The PSD model

The PSD model [18] is a recent conceptualization for designing, developing and evaluating persuasive systems. It consists of (a) the premises behind any persuasive system, (b) the persuasion context and (c) the persuasive software system features. Hence, according to the PSD model any persuasive system is based on eight premises detailed in [18] and listed here: P1: Useful; P2 User-friendly; P3: Unobtrusiveness; P4: Open for transparency; P5: Cognitive Consistency; P6: Incremental; P7: Information technology partiality; P8: Direct and indirect routes to persuasion.

The analysis of the *persuasion context* consists of looking into (1) the intent, (2) the event and (3) the strategy. The event comprises the use situation, user's characteristics, technological platform and environment. The strategy includes the message itself and the route to be used to achieve a goal.

The PSD model describes persuasive software system features grouped in four categories: (i) The **primary activity support category** focuses on supporting the activities that lead to achievement of the BCSS goals. These activities include

[1] Systems in which software interacts with other software, systems, devices, sensors and people [14].

reduction, tunneling, tailoring, personalization, self-monitoring, simulation and rehearsal. (ii) **Dialog support** refers to techniques/mechanisms to motivate users to use BCSS. This category includes praises, rewards, reminders, suggestions, similarity, liking and social role features. (iii) The **credibility category** relates to how to design a system so that it is more credible and thereby more persuasive. This category includes the following features: trustworthiness, expertise, surface credibility, real world feel, authority, third party endorsements, verifiability. iv) **The social influence category** describes how to design the system so that it motivates users by leveraging different aspects of social influence. The features that belong to this category are: social learning, social comparison, normative influence, social facilitation, cooperation, competition, recognition.

According to the PSD model, a behavior change can be divided into three categories:

- C-Change - or compliance change, is to make sure that the user complies with the request of the behavior change support system.

- B-Change - or behavior change, is to elicit a more enduring change than simply compliance a couple of times. Short time behavior change is easier to achieve than long-term behavior change.

- A-Change - or attitude change, is to influence the users' attitudes rather than behavior only. Changing the attitude of a user may be the most difficult type of change to achieve by a behavior change support system.

The outcomes of these C, B and A-Changes are the formation, alteration or reinforcement:

- F-Outcome - It means the formulation of a pattern in a situation where it did not exist before.

- A-Outcome - It means a change in the response of a user to an issue.

- R-Outcome – It means the reinforcement of current attitudes

2.2 The SQ model

The SQ model [9] is defined in terms of four *sustainability dimensions*: (i) *Technical dimension* addresses the long-term use of software-intensive systems and their appropriate evolution in an execution environment that continuously changes. (ii) *Economic dimension* focuses on preserving capital and (economic) value. (iii) *Social dimension* focuses on supporting current and future generations to have the same or greater access to social resources by pursuing generational equity. (iv) *Environmental dimension* aims at improving human welfare while protecting natural resources. For software-intensive systems, this dimension aims at addressing ecologic concerns, including energy efficiency and ecologic awareness creation.

Each dimension is characterized by a set of *Quality attributes*, which can be interdependent. Such dependency can be of two types: (i) it is *inter-dimensional* if it relates a pair of quality attributes defined simultaneously in two different dimensions (e.g. security defined in the *technical* dimension can influence security in the *social* dimension); and (ii) it is *intra-dimensional* if a dependency exists between two different quality requirements defined within the same dimension (e.g. in the *technical* dimension, security may depend on reliability).

Our SQ model supports the identification of sustainability design concerns, and the quality assessment of software architecture. The list of measurable attributes of the SQ model and corresponding contributions to the four dimensions can be found at [9], which has been empirically evaluated [19, 20].

3. NUX-based approach for deriving sustainability-quality requirements

In this section, we present the process needed for deriving sustainability-quality requirements from NUX results. As shown in **Figure 1**, the process consists of three stages: (i) UX assessment for understanding user needs, (ii) translating user needs into NFRs, and (iii) deriving sustainability-quality requirements from identified features and NFRs. The first two stages correspond to the NFR discovery approach proposed in [21], which uses the PSD model as a means to identify non-functional requirements (NFR) for a persuasive software system.

3.1 UX assessment for understanding user needs

UX assessment is supported by a wide range of research methods available, ranging from attitudinal evaluations (e.g. UX questionnaire, think-aloud) to behavioral evaluations (e.g. eye-tracking, activity trackers, emotions measurement). In this phase, in contrast to Sonnleitner et al. [22], we focus on negative User Experience (NUX) that is caused by the lack of fulfillment of needs during the interaction with a software product (a RSI software in our case). The effect of NUX impacts on the user attitudes ("what people say") and user behaviors ("what people do"). The outcome of this stage is the user feedback, information used for understanding user needs interacting with any persuasive software application.

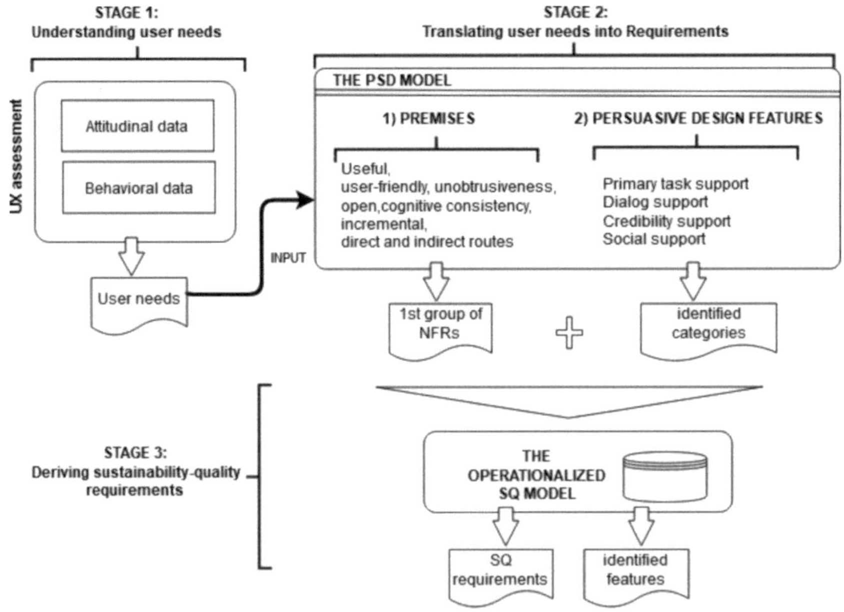

Figure 1.
Deriving sustainability quality requirements from NUX results.

3.2 Translating user needs into NFRs

In this stage, the user feedback from the UX assessment is used through two mapping steps to obtain a first group of NFRs and a set of categories of features respectively:

First step: by mapping the unfulfilled user needs (results) with the PSD model premises, we first discover a group of NFRs, which are directly discovered by translating the corresponding identified premises. Some examples of these NFRs are: usefulness (P1), cognitive consistency (P5), unobtrusiveness (P3), fun and enjoyment (P2), and trustworthiness (P4). Moreover, we also identify the affected categories of features described in the PSD model [18] (i.e., dialog support, primary activity support, perceived credibility, social influence).

Second step: by doing a content analysis, the unfulfilled user needs (expressed as user comments or suggestions) are categorized according to the identified categories.

3.3 Deriving sustainability-quality requirements

In order to facilitate the identification of quality requirements and features that contribute to the sustainability of software systems, a graph database has been created using the Neo4j Bloom tool. Such as shown in **Figure 2**, the data scheme includes elements from the SQ model (i.e., Dimension, Characteristic, Attribute) and the PSD model (i.e., feature and category), as well as the corresponding relationships among these elements. The relationships among the different elements (nodes) are represented by colored edges.

For instance, the edge COMPOSED_OF (green) is used for representing that a quality characteristic is defined in terms of a set of quality attributes of the SQ model. Similarly, there is a composition relationship between the features and categories of the PSD model (briefly introduced in Section 2.1).

Having the graph database, designers will be able to run (predefined) queries in the Neo4j browser. Query results are rendered either as a visual graph or a table format. **Figure 3** shows an example of a query result that displays all the quality attributes related to the usability characteristic and their corresponding contribution to the sustainability dimensions.

For illustrating the application of our approach for deriving sustainability quality requirements from a UX assessment, in the following section we present the design of a user study that aims to assess the experience of existing persuasive software applications for preventing RSI.

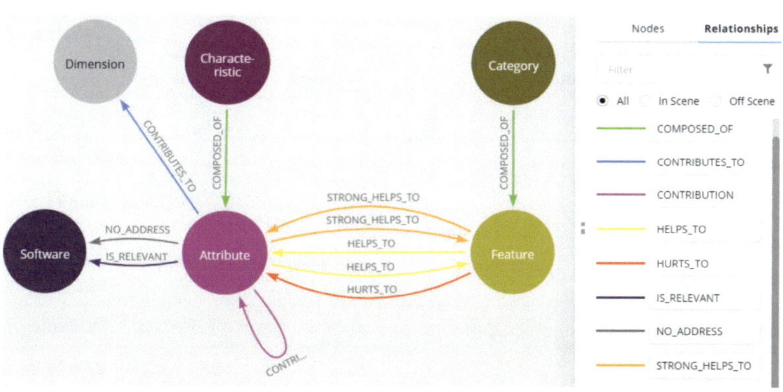

Figure 2.
Scheme with types of nodes and relationships in the graph database.

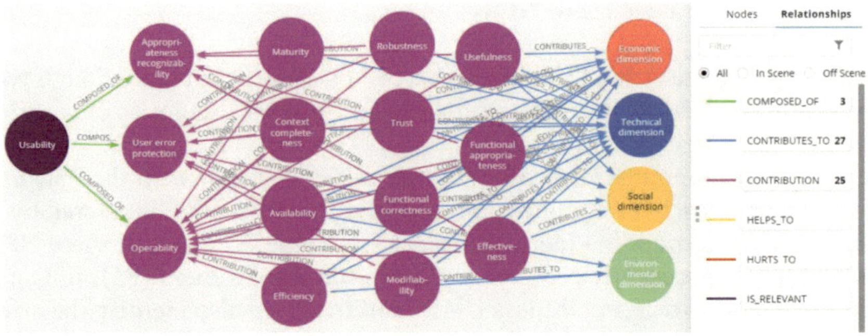

Figure 3.
Example of a query result in Neo4j: Quality attributes related to usability and their contributions to sustainability dimensions.

4. User study design

4.1 Participants

We targeted office employees working with computers (i.e. desktop, laptop) as RSI typically arises among this kind of user. A total of 30 people from four universities working in offices were invited to participate in the study. Twelve participants accepted our invitation. 3 were female and 9 male, whose age ranged from 21 to 45 years old.

4.2 Software and equipment

Table 1 shows the software available in the market, characterized by several commonplace functions, although their implementation can differ between systems:

1. Break reminder (BR) – Remind to take breaks based on several factors like elapsed time, how much/intensely you are working, natural rest patterns, times of day, etc.

2. Tracking (TK) – Track information like break-taking patterns, working hours per day, repetitions (e.g., keystrokes, mouse clicks).

3. Biofeedback (B) – Gain greater awareness of body functions primarily using instruments that provide information on the activity of those same systems.

Name	Features				Operative system
	Break reminder	Tracking	BioFeedback	Training	
Stretch, 2013	X	X		X	Windows
Workrave, 2013	X	X		X	Windows
Pokoy, 2014	X				Linux
Quick Pause, 2014	X	X			Windows
AntiRSI, 2016	X				Mac OS
SmartBreak, 2016	X	X	X		Windows, Mac OS
EyeLeo, 2016	X			X	Windows

Table 1.
Common functionalities of RSI software.

4. Training (TN) – Provide information on topics including workstation setup, body positioning, work-efficiency tips, psycho-social information, etc.

For practical reasons and given that participants self-reported to work with either Windows 10 Pro operating system or MAC OS, they were offered the two most complete options available, Workrave or SmartBreak, so that the study can run in their natural working environment. Although both applications, Workrave and SmartBreak, deliver reminders and enforce to take breaks, there are some differences as explained next.

Workrave is probably one of the most complete applications of its class. It considers micro-pauses, rest breaks, and guidance for exercise routines. This software is based on timers and keyboard/mouse activity, which determine when the actions must be displayed on screen. The user interface offers to configure a good number of parameters and provides a monitor on micro-breaks, rest breaks and working hour limit. The remarkable feature is the Training support, using an animated virtual human to demonstrate the exercises in addition to a textual description (see **Figure 4**), which could have some positive effects on attaining coaching goals [23].

SmartBreak has a minimal user interface consisting of a "stress" level bar, and an overlay message is displayed in the center of screen when a break is suggested. The unique feature supported by SmartBreak is the one on BioFeedback, trying to raise awareness on user's stress state, which in this particular application is based on keyboard and mouse usage only.

According to the PSD model, two types of behavior changes are addressed by Workrave and SmartBreak, as shown in **Table 2**.

The study was conducted in the natural settings of the subjects for 1 week (5 working days). Hence, they used their own computer and had to install by themselves the software application of their choice. The 8 subjects with a Windows computer

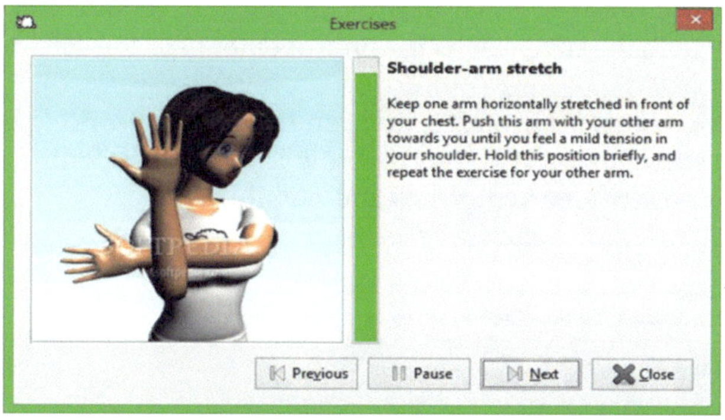

Figure 4.
Training coach demonstrating exercises to stretch body muscles in Workrave.

	B-change	A-change
F-Outcome	Workrave helps people to form healthier habits by doing stretching exercises	
A-Outcome		SmartBreak helps people to get awareness on their stress level determined based on the way how keyboard and mouse is used.

Table 2.
Outcome/changes analysis for Workrave and SmartBreak.

installed Workrave, whereas 4 users who had a laptop with Mac OS installed SmartBreak. Only one participant had used an RSI software in the past (i.e. Workrave).

4.3 Instrumentation

Participants were given the following instruments:

- An UX questionnaire that is based on the User Needs Questionnaire (UNeeQ) [22]. It is composed of two parts described as follows:

  User needs fulfillment: The first part of this questionnaire measures the user experience of a product or product concept based on the needs fulfillment. Given our assessment focus on BCSS for recovering and preventing RSI, we consider the premises of the PSD model for the formulation of 10 items regarding specific user needs (See **Table 3**). All the items were measured on a five-point rating scale (0–4) ranging from "not at all" to "highly".

  Positive and Negative UX: The second part of UNeeQ consists of six items regarding overall positive and negative UX measured also in a five-point rating scale (0–4). These overall UX items correspond to positive/negative emotions, feelings and experience. In our study with the purpose of avoiding confusions with the items regarding feelings, we decided to remove the items regarding emotions. This is because usually feelings and emotions are used interchangeably, but there are distinct differences between these two words. Feelings are mental associations, whereas emotions create reactions altering physical state. Emotions could be measured more objectively with techniques such as facial recognition, or monitoring physiological data (e.g. Skin conductance).

- Open questions to capture in free form text any additional issue regarding user experiences and feelings, either positive or negative, as well as to allow participants to make suggestions for improvements to the software applications.

- Demographic questions included age, gender, weekly working hours, and more specific questions that allowed us to know if our respondents suffered RSI in the past (i.e. identifying symptoms and possible triggers).

Premise	Items		Day1	Day5
P1	... I was doing something good for my body and mind.	I1	X	X
P2	... joy and pleasure.	I2	X	X
P4	... safe from uncertainties.	I3		X
P1, P6	... that I was more physically active (e.g. stretching exercises).	I4		X
P1, P6	... that I was less mentally tired	I5		X
P3, P2	... disturbed by the RSI software	I6		X
P4, P6	... that I was developing a deeper understanding on how to prevent RSI	I7	X	X
P5	... that I was acting according to my true interest (e.g. moving naturally to well-being)	I8		X
P5	... that the system was supporting me to make my commitments (e.g. changing work habits)	I9		X
P6	... a better sense of physical well-being.	I10		X

Table 3.
Adapted user needs questionnaire for RSI software.

4.4 Procedure

Each participant was asked to install one of the selected RSI software (Workrave or Smartbreak) in his/her own computer, and configure the timing parameters according to the given instructions.

Then, we asked them to use the RSI software while they were working with their computer for 1 week.

During the study, participants were allowed to adjust the values of any parameter (e.g. break times, sound of alarm) whenever they considered necessary. As the study was conducted in their natural working environment, they were allowed to abandon and not to finish the study. They were informed beforehand about the length of the study and the existence of a final questionnaire that should only be filled in if the study rules were met regarding duration and working with the computer normally.

Given UX can change over time, the data collection was carried out at two moments: (i) At the end of the first day, participants were asked to complete a first UX questionnaire. As some items of the UX questionnaire could not be experienced immediately by users, we considered only a subset of items for the first round listed in **Table 3**. (ii) At the end of the study, participants were asked to complete the second UX questionnaire.

5. Deriving sustainability quality requirements from UX assessment

In this section we present the results obtained following the NUX-based discovery process introduced in Section 3.

5.1 First stage: UX assessment for understanding user needs

According to our demographic data collected at the beginning of the study, we found that most of our participants worked more than 40 hours per week. The distribution is as follows: 5 subjects reported to work more than 45 hours per week. 5 subjects worked between 40 and 45 working hours. Only 2 subjects worked between 38 and 40 weekly hours.

Regarding RSI symptoms, most of the participants indicated felt fatigue, aching or shooting pain. 3 subjects did not experience none of the symptoms shown in **Table 4**. One of these 3 subjects decided to drop-out from the study after experiencing with the RSI software for 1 day.

Nine of our subjects considered stress as the main trigger of their RSI symptoms, followed by a bad ergonomic posture (8 subjects). Surprisingly, we found that omitting breaks and maximum exposure to technology were considered by less than half of the subjects that experienced RSI symptoms.

5.1.1 First impressions

At the beginning of the study we found that the RSI software was not enjoyable enough and pleasant. The most positive answers were given by 3 participants using Workrave, who indicated have enjoyed -more or less- with the software. However, despite this non-positive feeling, 7 of out 11 participants felt that using an RSI software was somewhat useful ("more or less" and "significantly high") to support their own wellness (body and mind). Over half of the participants reported that they understood how to prevent RSI during their first interaction with the software. **Figure 5** shows the frequency distribution of the responses in detail regarding these three items that concern about enjoyment, usefulness, and awareness respectively.

RSI symptoms	Responses
Aching or shooting pain.	50% (6)
Fatigue or lack of strength.	50% (6)
Weakness in the hands or forearms.	25% (3)
Tremors, clumsiness and numbness.	17% (2)
Chronically cold hands, particularly the fingertips.	17% (2)
Difficulty with normal activities like opening doors, turning on a tap.	8% (1)

Table 4.
RSI symptoms experienced by participants.

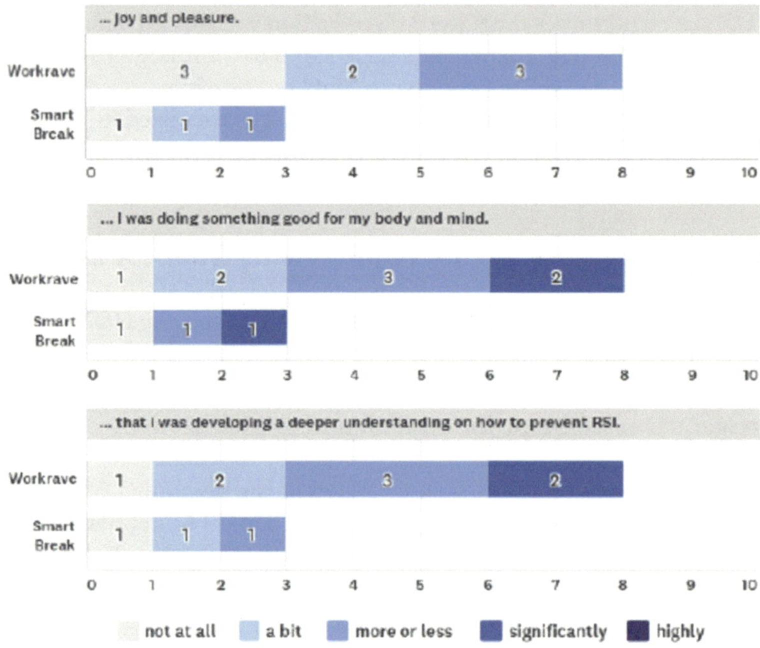

Figure 5.
User experience after day 1 (N = 11).

5.1.2 UX based on the need's fulfillment

A total of 9 participants completed the study and filled in the final questionnaire after day 5. **Table 5** shows the answers' frequency to the user experience questionnaire items and the weighted average. After several days of use, it is confirmed that the extent of "enjoyment and pleasant" is low. The "safe from uncertainties" score reinforces the idea that participants were not fully sure on how the break reminder and monitoring features work (e.g. how the breaks based on the stress level is determined by SmartBreak).

Most of the participants perceived that the software provided "a bit" support for items related to the physical and mental well-being. This feeling was confirmed by the last two items, where participants did not show to have the impression that the software was supporting their commitments and supporting behavior change in a natural way; therefore, more feedback from the software could be needed along with a different strategy to communicate how the software works and help to attain goals. Overall, it seemed that participants have got a moderate belief of "doing

Items	0	1	2	3	4	W-Avg
Joy and pleasure.	4	3	2	0	0	0.77
a better sense of physical well-being.	3	3	2	1	0	1.11
I was doing something good for my body and mind.	0	3	3	3	0	2
Safe from uncertainties.	6	0	2	1	0	0.77
I was more physically active (e.g. stretching exercises).	2	4	1	2	0	1.33
I was less mentally tired	4	3	1	1	0	0.89
Disturbed by the RSI software	0	4	2	1	2	2.11
I was developing a deeper understanding on how to prevent RSI	0	3	3	3	0	2
I was acting according to my true interest (e.g. moving naturally to well-being)	1	3	2	3	0	1.78
The system was supporting me to make my commitments (e.g. changing work habits)	2	3	2	2	0	1.44

Table 5.
UX based on the needs fulfillment on day 5 (N = 9).

something good for their body and mind", and have been developing some moderate understanding on how to prevent RSI.

5.1.3 UX variation analysis

In order to understand the possible variations of the user experience reported after the first day, we carried out an individual analysis with the participants (S1, S2, S5, S6, S9, S11, S12) who finished the study.

What stands out in **Table 6** is that there was not any variation in enjoyment and pleasant from day 1 to day 5. Similarly occur for the other two items concerning

Items		Day 1	Day5
joy and pleasure.	4		
	3		
	2	S1, S6	S6
	1	S2, S12	S1, S2, S12
	0	S5, S9, S11	S5, S9, S11
I was doing something good for my body and mind.	4		
	3	S1, S6	S2, S6, S9
	2	S5, S12	S1, S12
	1	S9	S5, S11
	0	S2, S11	
I was developing a deeper understanding on how to prevent RSI	4		
	3	S6	S2, S6, S9
	2	S1, S12	S1
	1	S2, S9	S5, S11, S12
	0	S5, S11	

Table 6.
User experience variations along the study.

usefulness and awareness. However, what we can remark is that there were two few positive variations (colored in green), which correspond to subjects S2 and S9. Both subjects started rating negatively, but at the end of the study, they considered that the software was good for their own wellness, and to understand better on RSI prevention. Both continued considering their level of enjoyment as negative though.

This was mainly because breaks used to be considered as interruptions. (S2) *"Interruptions while typing, and (interruptions) of the reasoning flow while working"*. But then these breaks were understood as a time for resting. (S9) *"It reminds me that I must relax"*. Unfortunately, this change of attitude occurred only in these two subjects.

We observed also a negative variation (colored in red) that was detected in subject S1, and subject S12. (S1) *"When focusing on a task, being required to take a break resulted to be from time to time counterproductive"*.

5.1.4 Subjective rating of positive and negative UX

Participants rated the four items of in second part of the UX questionnaire that correspond to positive/negative feelings and experience (i.e. "I had a positive experience", "I had a positive feeling", "I had a negative experience", "I had a negative feeling"). The collected responses at day 1 and day 5 (see **Appendix A**) have been processed to facilitate their assessment, leading to overall indicators of UX presented in **Tables 7** and **8**. **Table 7** depicts the frequency for both positive and negative UXs, in such a way that the overall UX is calculated by aggregating the experience and feeling single items. It reports the distribution of the overall UX at the beginning and at the end of the study. It clearly shows that there were a mix of positive and negative user experiences.

In order to calculate the overall UX scores, we averaged the values assigned by the respective participants using Workrave (WR) or SmartBreak (SB) along the study, which are depicted in **Table 8**. It shows the corresponding overall UX of both RSI software was somewhat negative.

On the overall positive UX, data revealed that software applications did not manage to raise a significant or very positive UX. At the end, the ratings for positive experiences suggest that systems were better considered than they were at the beginning, low to moderate though.

Overall negative UX frequencies suggest that at the beginning there were not much negative UX (notice that 46% replied *Not at all*), but participants manifested that negative experiences and feelings became more intense at the end of the study.

	Highly		More or less		Not at all	Not at all		More or less		Highly
At the beginning			**Negative**					**Positive**		
Workrave	9%	18%	9%	5%	32%	23%	14%	18%	9%	9%
Smart Break	0	0	9%	5%	14%	9%	9%	9%	0	0
Total	9%	18%	18%	10%	46%	32%	23%	27%	9%	9%
At the end										
Workrave	0	22%	17%	11%	17%	11%	11%	22%	22%	0
Smart Break	0	0	6%	28%	0	11%	0	17%	6%	0
Total	0	22%	22%	39%	17%	22%	11%	39%	28%	0

Table 7.
Frequency distribution for overall positive and negative UX.

	Day 1	Day5
Positive	1.4	1.73
Negative	2.68	2.51

Table 8.
Overall UX scores on day 1 and day 5.

5.2 Second stage: translating user needs into requirements

In order to discover requirements that should be addressed by a persuasive software application (RSI software in particular) designed for helping people change their behavior to achieve healthier habits, our analysis focuses on the UX results and open-ended descriptive user responses (comments and suggestions).

By mapping user needs (first step of second stage) that were not fulfilled and what people would like to have (results presented in Section 5.1), we found that the premises P1 (Useful), P2 (user friendly), P3 (Unobtrusiveness), P4(Open), P5 (Cognitive consistency), P6(Incremental) were affected, which were accordingly translated into usefulness, pleasure, unobtrusiveness, transparency, cognitive consistency, and awareness requirements. For instance, from the questionnaire (I4, I5) some participants showed their perceptions that the RSI software was not so useful for keeping them more physically active and less mentally tired. It is also important to remark that the usefulness perception of a software system could vary along the time. (i.e. at the beginning some users expected the system would help to change their habits positively but at the end this did not happen).

Analyzing these unfulfilled user needs, we found that both RSI software apps lack some features related to dialog support, primary activity support and credibility categories (second step of second stage). For example, **Table 9** shows the discovered requirements like transparency, awareness, and consistency, which are helpful for implementing features from the *dialog support* category (i.e., providing relevant, motivating and adequate feedback to its users). Other important requirements from the primary activity support category are pleasure, usefulness and unobtrusiveness. And transparency that helps also to the features from the credibility category.

5.3 Third stage: deriving sustainability-quality requirements

Considering the relations of the SQ model for each NFR(attribute) identified in the previous stage, we found out other relevant requirements that contribute to the sustainability dimensions.

For example, regarding the *primary activity support*, pleasure is positively affected by adaptability. This attribute is relevant for contributing to the social and technical sustainability of the RSI software. But as shown in **Table 9**, this requirement was not adequately addressed. If the RSI software had more information about the current situation, an adapted modality of breaks reminder could be delivered, which could influence positively on UX too.

It similarly occurred for usefulness, a key quality attribute that contributes to the economic, social and technical dimensions (see **Appendix B**), and it is positively related to learnability and effectiveness attributes (both from social dimension). From the UX assessment (**Table 9**), we corroborated that learnability and effectiveness were not satisfied by users. For instance, although Workrave provides an animated coach to teach a series of body movements, the UX could be affected if the content itself was not easy to learn/understand (learnability), or users were not able to change some of the exercises proposed by the system (tailorability).

Category of features	UNeeQ Item	User comments (C) and suggestions (S)	Premises	NFR
Dialog Support	I7, I10	C: "I cannot really say I may be satisfied with RSI prevention as I do not know about the state of my condition." S: "Give supporting messages that create awareness on RSI (why it is important)..."	P4, P6	**Transparency, Awareness**
	I8, I9	C: "the RSI software made me realize that I am frequently taking brakes even without tool support"	P5	**Cognitive Consistency**
Primary activity support	I2	C: "I was always annoyed by it, especially when it freezes the screen while I am in the middle of some work." C: "the RSI software resulted to be quite annoying when a break was triggered" C: "The animation starts the exercise but the instructions are in a long hard to read text, so at the beginning it is annoying to read and see how the animation started and you cannot follow it as still reading"	P2	**Pleasure** Usability (operability)
	I1, I4, I5	C: "It reminds me that I must relax".	P1, P6	**Usefulness,** Effectiveness
	I6	S: "..somehow having a do not disturb mode in which it does not interrupt you but uses a more subtly easy to inform you" C: "Interruptions while typing and (interruptions) of the reasoning flow while working."	P3, P2	**Unobtrusiveness**
	I6	C: "I was busy sending some important emails, and I had to wait for a couple of minutes for the computer to unlock." C: "I tried to follow the instructions of the break, but sometimes is not possible for me because you cannot interrupt your activity of you are completing something urgent or you are discussing our taking with someone else using the screen..."	P3	Timeliness
		S: "It would also nice to be able to choose your set of exercises, or changing them for time to time." S: "blocking interruption stressed me. So, I preferred the non-blocking interruption"	P6, P8	Tailorability
		C: "Suggested exercises are not clear and I had a small period of time to understand them"	P8	Learnability
		C: "The RSI software resulted to be quite annoying when a brake was triggered during a Skype call." C: "The software actually did not do anything. I did not get any alert, any break, etc. (I was following a course on coursera, which meant mostly watching a video) so maybe that is the reason."		Adaptability
Credibility	I3	C: "The system showed me a stress bar when I did not feel to be stressed"		Trust
		C: "I better understood the meaning of the countdown in the breaks (of the mouse is touched then it is stopped...). I had the feeling that when I skipped breaks, there were appearing often, and therefore interrupting my working activity is not nice."	P4	**Transparency**

Table 9.
NFR found from the UX assessment.

Given preferences can vary significantly over users with different profiles (ages, interests, etc.), tailorability is an important requirement that must also be considered. From user suggestions, we found that this requirement was not fully addressed. We also found that the break reminder could have a negative effect on

NFR	Sustainability Dimension	Candidate Features
Usefulness	Economic, Social and Technical	Self-monitoring (HELPS_TO–>)
Unobtrusiveness	Social	Tunneling and Reduction (<−-HELPS_TO)
Timeliness	Social	Rehearsal and Tunneling (<−-HELPS_TO)
Tailorability	Social and Technical	Tailoring and Personalization (<−-HELPS_TO)
Transparency	Social	Suggestion, Real-world feel, Praise, Verifiability, Authority, Reminders, Third-party endorsements, Trustworthiness, Rewards, Expertise, Surface credibility (<−-HELPS_TO)

Table 10.
Example of candidate features and sustainability dimension covered by discovered NFR.

the user's experience due to that this reminder sometimes occurred in a not favorable time (lack of timeliness).

Finally, the third category, *credibility*, consists of two requirements that relates to the confidence on software systems that behave as intended (trust) and transparency of the system (implications and consequences of functionality such as skipping or postponing breaks or when they will be offered should be clear to the user regardless of the internal details on how they are being handled or implemented). Trust, an attribute from the social and technical dimension of the SQ model, was derived thanks to the positive relation with the pleasure attribute from the social dimension.

Table 9 presents the 6 NFRs discovered at the second stage (text in bold in the last column) and 7 NFR derived from the SQ model at the third stage. By means of our approach, in this stage we can also determine (i) the new potential features that should be considered in further versions of the software for RSI prevention, and the sustainability dimensions that can be covered with the implementation of the discovered NFRs. Some of the identified features as potential improvements are shown in **Table 10**. For example, the self-monitoring feature might help to address the usefulness of applications because the system would be able to provide other means to track user status through self-monitoring.

6. Threats to validity

In this section, we discuss the threats to the validity of our user study.

6.1 Internal validity

A threat related to the *instrumentation* could be caused by the questionnaires used during the study. Aimed at mitigating this threat, the online questionnaires were carefully reviewed and tested before running the user study. Given the type of collected data (e.g. actual number of working hours), we decided to make responses anonymous (IP were not recorded) and participants were informed of such anonymity. But for our study, responses traceability between the first and second questionnaire was required. Thus we asked participants to create a fictitious username that still kept their responses anonymous.

Considering our user study took five working days and it was conducted in a real setting with a null control, the threat of having a high number of dropouts (*mortality*) could not be reduced. But we tried to reduce it, by informing to the potential participants beforehand about the length of the study, and the existence of a final questionnaire that should only be filled in if the study rules were met, regarding duration and having worked with the computer normally during that time. We got twelve participants from thirty invitees, who accepted our invitation. Once the study was running we sent a couple of emails to remind deadlines for completing the respective UX questionnaires. At the beginning of the study we had one participant who dropped out from the study. And two more at the end of the study.

6.2 External validity

The first threat to consider is concerned with the *interaction of selection and treatment*. This is the effect of having non-representative subjects. We attempt to mitigate this threat by inviting office employees working with computers (i.e., senior researchers, programmers), as potential users interested in preventing RSI. Regarding the size of our sample (2 partial responses, and 9 full responses), it could be considered as small. But according to Bevan et al. [24], 80% of usability findings are discovered after five participants. It is important to point out that our user study focuses on the user experience and not on the effectiveness of the RSI software for changing the behavior, and therefore we consider that the size of our sample is still good enough for illustrating the application of our NUX-based approach. Indeed, if the focus were on the behavior change effects, studies would necessarily be longer.

The second threat is concerned with the *interaction of setting and treatment*. We mitigate this threat by conducting the user study in their natural working office environment. Although originally our data collection plan was for 2 weeks, the study was executed only for 1 week due to negative experiences already experienced during the first 5 working days. Another threat is related to the representativeness of the selected experimental objects (RSI software), although we focus on two desktop applications, they cover common features of RSI software. However, we cannot generalize the results to any persuasive software system since according to the PSD model, the social influence category was not covered. Thus, this limitation can be mitigated by means of further replications including other types of persuasive software systems (e.g. activity trackers to encourage physical activity in non-working environments at leisure time) where other requirements could be discovered.

7. Conclusions

The present research presents an approach for deriving sustainability quality requirements from negative user experience. It starts by understanding the fulfillment of user needs through a UX assessment (first stage). We used the PSD model as the theoretical framework for designing our empirical research of persuasive systems. A user study with 12 subjects working on their natural office environment was carried out. Our UX assessment focused on two popular software systems for preventing RSI (i.e. Workrave and Smartbreak). This study revealed that generally most of the subjects had a negative UX with both RSI software at the end of the study. It is important to remark that from our UX variation analysis, UX of the participants was not much negative at the beginning (46% replied "Not at all"), but along the study their experiences and feelings were changing to be more negative.

From the UX results based on needs fulfillment and user comments/suggestions, 6 NFRs of RSI software (i.e., usefulness, pleasure, unobtrusiveness, transparency, cognitive consistency, and awareness) were discovered (second stage). Then 7 additional NFRs were derived from the SQ model by means of the existing predefined relations among quality attributes that contribute to the social, technical, economic, and environmental dimensions (third stage). All these requirements are helpful for identifying candidate features related to dialog support, primary activity support and credibility categories. Addressing requirements such as awareness, pleasure and consistency, RSI software apps will be in a better capability to provide relevant, motivating and adequate feedback to users. The second group of requirements such as unobtrusive, timely, learnability, usability (operability), usefulness, adaptability will enable RSI software apps to provide a better activity support for the system goal achievement (i.e. reduction and prevention of RSI, main goal of our selected software). The third group of requirements relates to the perceived credibility of the software system, where transparency and trust are very important requirements that positively could affect users to continue using the software system.

In order to support the derivation of requirements and features for improving the sustainability of persuasive systems (RSI software apps in our case), our SQ model along with its relationships with the PSD model has been implemented in Neo4j Graph platform, allowing us to search for potentially relevant attributes by querying and navigating the model interactively.

Appendix A: data collection

At the beginning	0	1	2	3	4
I had a positive experience					
Workrave	3	1	2	1	1
Smart Break	1	1	1	0	0
I had a positive feeling					
Workrave	2	2	2	1	1
Smart Break	1	1	1	0	0
I had a negative experience					
Workrave	3	1	1	2	1
Smart Break	2	0	1	0	0
I had a negative feeling					
Workrave	4	0	1	2	1
Smart Break	1	1	1	0	0

Table 11.
Overall UX at the beginning of the study (N = 11). Data collected from the RSI case presented in [7].

At the end	0	1	2	3	4
I had a positive experience					
Workrave	1	2	1	2	0
Smart Break	1	0	1	1	0
I had a positive feeling					
Workrave	1	0	3	2	0
Smart Break	1	0	2	0	0
I had a negative experience					
Workrave	2	1	1	2	0
Smart Break	0	3	0	0	0

At the end	0	1	2	3	4
I had a negative feeling					
Workrave	1	1	2	2	0
Smart Break	0	2	1	0	0

Table 12.
Overall UX at the end of the study (N = 9). Data collected from the RSI case presented in [7].

Appendix B: the SQ model

Quality attributes (QA)		Generic definition	TECH	ENV	ECON	SOC
Compatibility	Co-existence	Product can perform its functions efficiently while sharing environment and resources with other products	■	■		■
	Interoperability	a system can exchange information with other systems and use the information	■			■
Context coverage	Context completeness	system can be used in all the specified contexts of use	■		■	
	Flexibility	system can be used in contexts beyond those initially specified in the requirements.			■	
Effectiveness	Effectiveness	accuracy and completeness with which users achieve specified goals.	■		■	■
Efficiency	Efficiency	resources expended in relation to the accuracy and completeness with which users achieve goals	■	■	■	
Freedom from risk	Economic risk mitigation	system mitigates the potential risk to financial status in the intended contexts of use			■	
	Environmental risk mitigation	system mitigates the potential risk to property or the environment in the intended contexts of use.		■		■
	Health and safety risk mitigation	system mitigates the potential risk to people in the intended contexts of use.				■
Functional suitability	Functional appropriateness	the functions facilitate the accomplishment of specified tasks and objectives.	■		■	
	Functional correctness	system provides the correct results with the needed degree of precision.	■		■	
	Functional completeness	degree to which the set of functions covers all the specified tasks and user objectives.			■	
Maintainability	Modifiability	system can be effectively and efficiently modified without introducing defects or degrading existing product quality.	■	■	■	
	Modularity	system is composed of components such that a change to one component has minimal impact on other components.	■			
	Reusability	an asset can be used in more than one system, or in building other assets.		■	■	
	Testability	effectiveness and efficiency with which test criteria can be established for a system.	■			
Performance efficiency	Capacity	the maximum limits of a product or system parameter meet requirements.	■			
	Resource utilization	the amounts and types of resources used by a system, when performing its functions, meet requirements.		■		
	Time behaviour	response, processing times and throughput rates of a system, when performing its functions, meet requirements.	■	■		
Portability	Adaptability	system can effectively and efficiently be adapted for different or evolving hardware, software or usage environments.	■			
	Replaceability	product can be replaced by another specified software product for the same purpose in the same environment.			■	
Reliability	Availability	system is operational and accessible when required for use.	■	■	■	■
	Fault tolerance	system operates as intended despite the presence of hardware or software faults.	■			
	Maturity	system meets needs for reliability under normal operation.	■			
	Recoverability	system can recover data affected and re-establish the desired state of the system is case of an interruption or a failure.	■		■	
Satisfaction	Trust	stakeholders has confidence that a product or system will behave as intended.	■		■	■
	Usefulness	user is satisfied with their perceived achievement of pragmatic goals.	■		■	■
Security	Accountability	actions of an entity can be traced uniquely to the entity.				■
	Authenticity	the identity of a subject or resource can be proved to be the one claimed.				■
	Confidentiality	system ensures that data are accessible only to those authorized to have access.				■
	Integrity	system prevents unauthorized access to, or modification of, computer programs or data.	■			■
Usability	Appropriateness recognizability	users can recognize whether a system is appropriate for their needs, even before it is implemented.				■
	Learnability	system can be used to achieve specified goals of learning to use the system.			■	■
	Operability	system has attributes that make it easy to operate and control.				■
	User error protection	system protects users against making errors.	■		■	■
Accessibility	Accessibility	system can be used by people with the widest range of characteristics and capabilities.				■
Robustness	Robustness	Refers to the capability of the system to behave in an acceptable way in unexpected situations	■			
Survivability	Survivability	The degree to which a system continues to fulfil its mission by providing essential services in a timely manner in spite of the presence of attacks	■			
Data Privacy	Data Privacy	privacy concerns arise wherever personally identifiable information is collected, stored, or used				■
Timeliness	Timeliness	the fact or quality of being done or occurring at a favourable or useful time				■
Reguation compliance	Reguation compliance	Allows to draw conclusions about how well software adheres to application related regulations in laws.		■		■
Scalability	Scalability	the ability of a computing process to be used or produced in a range of capabilities.			■	
Tailorability	Tailorability	system's capability to allow users to create or enable new configuration of functionality as well as control information provision.	■			■
Unobtrusiveness	Unobtrusiveness	system should avoid disturbing users while they are performing their primary tasks with the aid of the system.				■
Transparency	Transparency	the degree to which stakeholders can answer their questions by using the information they obtain about a software system during its life cycle.				■

Figure 6.
Quality attributes of the SQ model defined according to [9, 19].

Author details

Nelly Condori-Fernandez[1,2*], Marcela Quispe-Cruz[3], Alejandro Catala[4], Joao Araujo[5] and Patricia Lago[2,6]

1 Database Laboratory, University of A Coruña, A Coruña, Spain

2 Vrije Universiteit Amsterdam, The Netherlands

3 National University of Saint Agustin, Arequipa, Perú

4 Centro Singular de Investigación en Tecnoloxías Intelixentes (CiTIUS), Universidade de Santiago de Compostela, Spain

5 Universidade Nova de Lisboa, Portugal

6 Chalmers University of Technology, Sweden

*Address all correspondence to: n.condori.fernandez@udc.es and n.condori-fernandez@vu.nl

References

[1] Fogg, B. Persuasive Technology. Using Computers to Change What We Think and Do. CA, USA: Morgan Kaufmann Publishers Inc. San Francisco; 2003.

[2] Froehlich, J. Promoting energy efficient behaviors in the home through feedback: The role of human computer interaction. In: HCIC 2009 Winter Workshop

[3] Prochaska, JO, DiClemente, CC. The transtheoretical approach. Administration and Policy in Mental Health and Mental Health Services Research. 2005; 147–171. DOI: 10.1093/med:psych/9780195165791.003.0007

[4] Locke, E, Latham, G. A theory of goal setting & task performance. The Academy of Management Review. 1991; 16: 480–483

[5] Fogg, B. A behavior model for persuasive design. Proceedings of the 4th International Conference on Persuasive Technology - Persuasive '09. 2009; 1

[6] Harjumaa, M, Oinas-Kukkonen, H. An analysis of the persuasiveness of smoking cessation websites. In: The Second International Symposium on Medical Information and Communication Technology; 2007

[7] Condori-Fernández, N, Bolos, AC, Lago, P. Discovering requirements of behaviour change software systems from negative user experience. In: Proceedings of the 40th International Conference on Software Engineering: Companion Proceedings (ICSE 2018); May 27–June 03 2018; Gothenburg, Sweden; 2018. p. 222–223

[8] Ameller, D, Franch, X, Gómez, C, Martínez-Fernández, S, Araujo, J, Biffl, S, Cabot, J, Cortellessa, V, Méndez, D, Moreira, A, Muccini, H, Vallecillo, A, Wimmer, M, Amaral, V, Bühm, W, Bruneliere, H, Burgueño, L, Goulao, M, Teufl, S, Berardinelli, L. Dealing with non-functional requirements in model-driven development: A survey. IEEE Transactions on Software Engineering. 2019; 1–1. DOI: 10.1109/TSE.2019.2904476

[9] Condori-Fernandez, N, Lago, P. Characterizing the contribution of quality requirements to software sustainability. Journal of systems and software. 2018; 137: 289–305. DOI: 10.1016/j.jss.2017.12.005

[10] Calero, C, Moraga, MA, Bertoa, MF. Towards a software product sustainability model. CoRR. 2013; abs/1309.1640

[11] Venters, C, Jay, C, Lau, L, Griffiths, MK, Holmes, V, Ward, R, Austin, J, Dibsdale, CE, Xu, J. Software sustainability: The modern tower of babel. In: RE4SuSy: Third International Workshop on Requirements Engineering for Sustainable Systems. 2014

[12] Lago, P, Kocak, SA, Crnkovic, I, Penzenstadler, B. Framing sustainability as a property of software quality. Communications of the ACM; 2015. 58 (10): 70–78

[13] Brundtland, G, Khalid, M, Agnelli, S, Al-Athel, S, Chidzero, B, Fadika, L, Hauff, V, Lang, I, Shijun, M, Morino de Botero, M, Singh, M, Okita, S. Our common future (brundtland report). Technical report, World Commission on Environment and Development; 1987.

[14] Hölzl, M, Rauschmayer, A, Wirsing, M. Engineering of software-intensive systems: State of the art and research challenges. In Wirsing, M., Banâtre, J.P., Hölzl, M., Rauschmayer, A., eds.: Software-Intensive Systems and New Computing Paradigms. Springer-Verlag, Berlin, Heidelberg. 2008; 1–44

[15] Becker, C, et al. Requirements: The key to sustainability. IEEE Software. 2016; 33(1): 56–65

[16] Assefa G., Frostell B., Social sustainability and social acceptance in technology assessment: A case study of energy technologies, Technology in Society, Volume 29, Issue 1, 2007, Pages 63–78.

[17] Condori-Fernandez N., "HAPPYNESS: an emotion-aware QoS assurance framework for enhancing user experience," 2017 IEEE/ACM 39th International Conference on Software Engineering Companion (ICSE-C), Buenos Aires, 2017, pp. 235–237, doi: 10.1109/ICSE-C.2017.137.

[18] Oinas-Kukkonen, H, Harjumaa, M. Persuasive systems design: Key issues, process model, and system features. Communications of the Association for Information Systems. 2009; 24: 485–500

[19] Condori-Fernandez N, Lago P, Luaces MR, Places ÁS. An Action Research for Improving the Sustainability Assessment Framework Instruments. *Sustainability*. 2020; 12(4): 1682.

[20] Condori-Fernández N., Lago P., Towards a Software Sustainability-Quality Model: Insights from a Multi-Case Study. 1–11 2019 RCIS https://doi.org/10.1109/RCIS.2019.8877084

[21] Condori-Fernández, N, Araújo, J, Catalá, A, Lago, P. Towards a non-functional requirements discovery approach for persuasive systems. 2020 SAC; 2020. p. 1418–1420. DOI: 10.1145/3341105.3374059

[22] Sonneleitner, A, Pawlowski, M, Kässer, T, Peissner, M. Experimentally Manipulating Positive User Experience Based on the Fulfilment of User Needs. 14th International Conference on Human-Computer Interaction (INTERACT); Sep 2013; Cape Town, South Africa: Springer, LNCS-8120 (Part IV); 2013. p.555–562

[23] Bickmore, TW, Silliman, RA, Nelson, K., Cheng, DM, Winter, M, Henault, L., Paasche-Orlow, MK. A randomized controlled trial of an automated exercise coach for older adults. J Am Geriatr Soc. 2013 Oct; 61 (10):1676–83. DOI: 10.1111/jgs.12449.

[24] Bevan N., Barnum C., Cockton G., Nielsen J., Spool J., and Wixon D., 2003. The "magic number 5": is it enough for web testing? In CHI '03 Extended Abstracts on Human Factors in Computing Systems (CHI EA '03). ACM, New York, USA, 698–699

Chapter 7

Implementing Visual Analytics Pipelines with Simulation Data

Taimur Khan, Syed Samad Shakeel, Afzal Gul, Hamza Masud and Achim Ebert

Abstract

Visual analytics has been widely studied in the past decade both in academia and industry to improve data exploration, minimize the overall cost, and improve data analysis. In this chapter, we explore the idea of visual analytics in the context of simulation data. This would then provide us with the capability to not only explore our data visually but also to apply machine learning models in order to answer high-level questions with respect to scheduling, choosing optimal simulation parameters, finding correlations, etc. More specifically, we examine state-of-the-art tools to be able to perform these above-mentioned tasks. Further, to test and validate our methodology we followed the human-centered design process to build a prototype tool called ViDAS (**Vi**sual **D**ata **A**nalytics of **S**imulated Data). Our preliminary evaluation study illustrates the intuitiveness and ease-of-use of our approach with regards to visual analysis of simulated data.

Keywords: visual analytics, machine learning, interaction, user experience, assistive technologies

1. Introduction

This section gives a brief overview of this chapter. It starts by discussing what usability is and why advanced analysis is required to extract useful information from raw data. Later, the idea and need for visual analytics is discussed and its implementation pipeline is presented which shapes the basic working structure of Visual Analytics of Simulated Data (ViDAS). Furthermore, the context of ViDAS is presented, keeping the human-centered design in mind. While the expert evaluation and feedback is based on simulation data, ViDAS is equivalently capable of handling usability data.

1.1 Motivation

With the technological advancements, data is automatically recorded and stored using various sensors and monitoring systems. This large amount of complex data is known as big data which is unstructured and contains hidden values. Therefore, there is a need to analyze this data, discover new values, and gain an in-depth understanding to efficiently manage and organize it [1].

Currently, there are a number of tools available in the market that help its users in analyzing the data and finding trends in it. These tools are mainly divided into two main categories; DM (Data Mining) tools and BI (Business Intelligence) tools. The DM tools are mainly focused on applying advanced machine learning models to the data and do not incorporate the advanced visualization techniques to go with it. The BI tools are mainly focused on EDA (Exploratory Data Analysis) techniques and include various clustering techniques along with interactive visualizations that help in understanding the data. However, the BI tools do not include built-in machine learning models and instead, rely on either third-party extensions or ask the user for a premium subscription.

This chapter presents a solution in the form of a web-based tool called ViDAS that incorporates the best features from both DM and BI tools. ViDAS combines these features in such a way that makes it effortless for the user to apply machine learning techniques on a complex data set, and then visualize the transformed data interactively.

1.2 Background

Intuitive computing technologies make their way into daily life and at the same time, the market is saturated with rival brands. This has made usability more popular in recent years, as businesses see the advantages of researching and designing their products using user-oriented approaches rather than traditional methods. Through knowing and studying the relationship between the product and the customer, the usability specialist may also have perspectives that are unfeasible by traditional market research. For example, after examining and evaluating customers, the usability specialist could recognize the requisite features or design shortcomings that were not expected.

Usability can be defined as the capacity of a software system to provide a condition for its users to perform tasks in a safe, effective, and efficient manner while enjoying user experience [2]. Usability requires techniques for assessing it, such as needs analysis [3] and the study of the values underlying perceived utility or beauty of the object. In the field of human-computer interaction and computer science, usability studies the sophistication and consistency with which interaction with a software system is built. Usability finds customer satisfaction and utility to be a quality component and strives to improve user experience through iterative design. Different researchers often focus on different parameters of usability, usability consultant Jakob Nielsen and computer science professor Ben Schneiderman have written (separately) about a framework of system acceptability, where usability is a part of "usefulness" and is composed of: Learnability, Efficiency, Memorability, Errors, and Satisfaction [4].

There are various methods that allow data collection of the above-mentioned parameters such as in-depth evaluations with a focus group, documenting the user experience, etc. Normally these parameters are analyzed using MS Excel and other traditional methods which require a lot of manual work. This is where the concept of visual analytics comes in as it provides an alternative approach that greatly supports the analysis process by the use of machine learning methods and detailed visualizations. Additionally, some of this usability data maybe complimented through simulations or machine learning methods that can be then analyzed to support the user-testing process.

In most cases, the dependencies and correlations of these parameters are not clearly identifiable, which forces the data analyst to make an educated guess. This guess is solely based on the expertise of the analyst which may result in extra time and effort spent in testing the focused parameters. This approach is known as the "trial-and-error approach" [5] which focuses on finding a good solution and states that the data analyst must spend more time examining the parameters than building the model. In such a scenario, a better approach is to use visual analytics to understand the data better and find hidden relationships between the parameters. Visual analytics aims to help data analysts in identifying correlated parameters than relying on just the trial-and-error approach.

1.2.1 Visual analytics

Visual analytics is a human-centric process that combines techniques from graphics, visualization, interaction, data analysis, and data mining to support reasoning and sense-making for complex problems and extract relevant information from the raw data. While simple visualization techniques can be applied to the simulation data to investigate different parameters, find patterns, and visualize dependencies, data mining and machine learning helps with examining the data further through techniques such as forecasting, clustering, regression, etc. Visual Analytics is comprised of two parts; data analytics and data visualization. These two approaches co-exist and support the visual analytics process to understand complex data and find patterns in it.

Furthermore, visual analytics goes a step further and includes the approach of human-in-the-loop [6]. This approach combines the perceptual capabilities of a human mind along with the interactive data visualizations to apply visual analytics. Visual analytics is not a tool, but a human-centric process that aims at integrating human perception in to the visual data exploration process. It requires the specialist to first understand the data and its context. After the data is prepared, interactive visualizations can be used to find patterns and extract useful information from the data. An ideal visual analytics solution should require little to no coding, allow the user to combine from multiple sources, offer easy-to-customize interactive visualizations, include the feature to drill down the data at any level of detail, and combine multiple views to get an overall understanding of the data.

Currently, there is a lot of work being done in academia and industry towards visual analytics solutions to assist in the sense-making of the data [7]. There are a number of commercial business intelligence solutions that specialize in data discovery such as Tableau [8], Qlik Sense [9], Power BI [10], etc. Additionally, a number of data mining tools are available such as KNIME [11], RapidMiner [12], Orange [13], Weka [14], KEEL [15], etc. that focus on applying machine learning models and provide visualizations to help understand the raw data and find a pattern in it. However, there is a lack of tools that perform both data discovery and apply machine learning models.

1.2.2 Visual analytics pipeline

To apply visual analytics for both research and industrial applications, an appropriate definition and implementation of visual analytics pipeline must be followed that provides an effective abstraction for designing and implementing visual analytics systems [16]. The most common visual analytics pipeline can be seen in **Figure 1**. This conventional pipeline guides the visual analytics processes as an abstract outline which includes four major procedures and the relationships between them. This subsection explores these procedures in detail and discusses their role in the visual analytics pipeline.

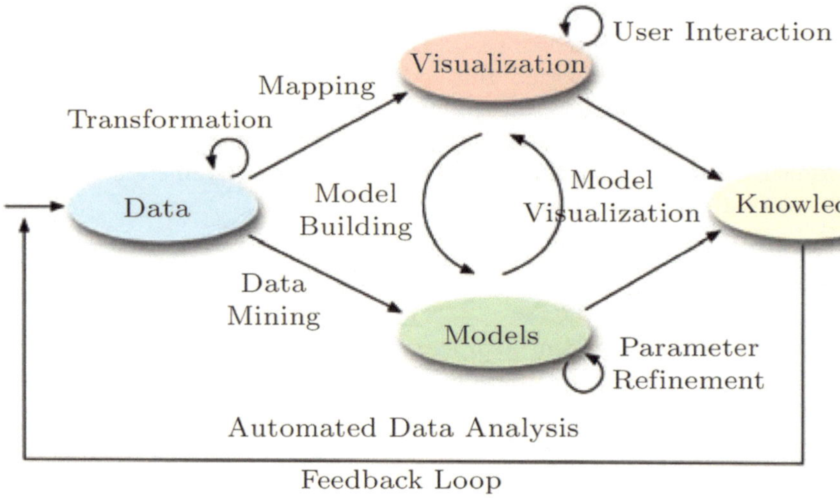

Figure 1.
Visual analytics pipeline by Keim et al. [17].

Data involves all the steps that are required to prepare the data set for visualizations and data analytics. These steps mainly include preparing and cleaning of data due to it being noisy and having missing values. All of these steps are collectively known as preprocessing steps and including data cleaning, data integration, data transformation, data reduction, and data discretization.

Visualization refers to the visual representation of data where the focus is on producing images that communicate the relationships between different attributes. This is achieved through the use of systematic mapping that establishes how the data values will be represented visually, determining how and to what extent a graphic mark property such as size or color will change to reflect the changes in the value of data. Effective visualizations help users to analyze the data and make complex data sets more accessible, understandable, and usable.

Model refers to the data analysis or machine learning methods that are applied to extract information out of the data which can later be visualized for crucial knowledge. These methods may include the steps from EDA in combination with machine learning algorithms such as Statistical models or Clustering models. After these methods have been applied, the resultant data is required to be visualized to extract knowledge that can help in understanding the data.

Knowledge refers to the process of generating a conclusion which either accepts or rejects a hypothesis. It is not a procedure by itself but an end result in the visual analytics pipeline. Knowledge is extracted from the data by applying all the above-mentioned procedures in order to understand the final visualization and get some insight into the data.

2. Related work

The relationship among all the procedures can be seen in the visual analytics pipeline i.e. visual mapping, data mining, model building, and model visualization. This section examines the different tools or techniques one would need to be able to perform the visual analytics pipeline tasks. As such Mapping and Model Visualization are grouped together under visualization tools while Transformation, Data Mining, and Model Building are grouped together under data mining tools. As shown in the visual analytics pipeline, Visual Mapping is the relationship between Data and Visualization whereas Model Visualization is the relationship between Model and Visualization. In both cases, one would need to rely on data mining tools that apply machine learning models to the data and transform it. Additionally, visualization tools are required that focus on visualizing the data and model respectively.

2.1 Visualization tools

In contrast to open-source libraries for creating charts, there are also configurable (drag and drop) tools available for data visualization. In this context, we shall explore Tableau and Qlik Sense which are the current market leaders in data visualization and data presentation. These tools make it easier to extract and convey key patterns and insights, something that is important in both the Mapping and Model Visualization relations.

Tableau is one of the first tools we consider when we talk about commercial data visualization tools. It has connectivity to multiple data sources, many different switchable chart formats, and a sophisticated mapping capability that can easily convert simple Excel data to colorful dashboards with a lot of interactivity [8]. It has a step-by-step configurable interface from creating charts in sheets to filtering and combining multiple charts that form a Dashboard to overall storytelling.

Qlik Sense is a commercial data visualization and analytics tool that enables the user to import and aggregate data from diverse sources and use the data visualization tools of the software to convert raw data into meaningful information. It has an in-memory data storage engine which helps in dynamic visualization building [9]. Qlik Sense also follows a step by step procedure such as Tableau where each sheet (which is also a dashboard) may contain multiple charts; the sheets are then used to create a story by adding a snapshot of the charts or the complete sheet to the storyline.

2.2 Data mining tools

Another approach is to use Data Mining Tools in order to assist with the Data Transformation, Data Mining, and Model Building relations. There are various data mining tools available which are widely used by organizations for data transformation, data mining, and model building such as KNIME, RapidMiner, KEEL, Orange, WEKA, and many more. This subsection only discusses KNIME and RapidMiner as they are the most widely used data mining tools currently in the market.

KNIME [11] is an open-source data analytics tool that is written in Java and is based on the Eclipse platform. KNIME provides a user-friendly workspace and is based on the idea of graphical workflows to design the data analytics pipeline. It provides hundreds of nodes that incorporate data I/O, data cleaning, data manipulation, Machine Learning (ML) methods, scripting, and visualizations that can easily be used to create a workflow using a simple drag-and-drop approach [18].

RapidMiner [12] is an open-source tool that integrates a number of packages including text mining, ML, predictive analysis, DM, and business analytics [18]. Based on the tool, a desktop software is developed which is known as RapidMiner Studio which provides a GUI. With RapidMiner Studio, the user can perform DM and predictive tasks by creating workflows and then visualizing the output in an interactive representation [19]. RapidMiner allows scripting as well as workflows and is constantly being updated.

2.3 Summary state of the art

While all the tools mentioned above have their pros and cons, none of them completely cover the data analytics pipeline. The business intelligence tools are more inclined towards interactive visualizations and presentation of data, while the data mining tools are more focused on applying machine learning models to data and transforming it. ViDAS fills this void by combining the features of these tools and accommodating the complete visual data analytics pipeline.

3. User-context and requirements

Identification of user-context is the first step in a project based on human-centered design. It refers to understanding the users and identifying the intended way the project will be used by those users. It could be sufficient to just identify the stakeholders, but in most cases identifying the purpose and scope of the project help recognize the environment, it will be used in. Similarly in ViDAS development, the first step consisted of doing background research on the stakeholders, understanding their needs, and documenting the requirements. It was found that the stakeholders have a simulation model which generates data based on individual simulation runs. This data can not be easily understood by just looking at its tabular form, which is why advanced analytics is needed that extracts crucial information from the raw data and visualizes it in order to gather the knowledge. Once the context of ViDAS was understood, the next step was to gather the requirements for the development of the tool.

After identifying and understanding the context of ViDAS in terms of human-centered design, the next step is to gather and specify the requirements which will be the basis for the development and evaluation phases. The requirement gathering process is not as simple as discussing the needs of the stakeholders and documenting them. Instead, it includes making the stakeholders realize which requirements are needed according to the scope and context of the project [20]. The requirements of a project are further divided into three main categories; Business Requirements, User Requirements, and System Requirements [21]. In this section, we will focus on the business, user, and system requirements of ViDAS and discuss them in detail.

3.1 Business requirements

Business Requirements consist of the high-level requirements that answer generic questions to define the overall scope of the project [22]. These requirements also include the stakeholder's objectives and the needs of the target users for which the system is to be developed. The basic requirements were gathered from the initial communication with the stakeholders and stated that the users of ViDAS will be software engineers that will possess some basic knowledge of data analytics and visualization. The users will get the data from their clients and provide them with useful insights on raw, unprocessed data. For ViDAS, the business requirements specified that the user should be able to upload raw data into the tool, pre-process and transform the data, apply machine learning models, and visualize the data to extract useful knowledge. In addition to this, the user should also be able to get a general overview of the complete data set in the form of a visualization. The business requirements for ViDAS have been summarized in **Table 1**.

ViDAS Business Requirements
Apply data analytics on raw data
Visualize the data to get an overview
Apply machine learning models for data analytics
Pre-process and transform data according to the user's requirements
Visualize the transformed data

Table 1.
Summary of ViDAS business requirements.

3.2 User requirements

User Requirements are the specifications of how the user wants to complete certain tasks which are based on the business requirements of the project. User requirements include designing the layout of the system, the sitemap, and developing prototypes while keeping all the user goals in mind. With the help of user requirements, the user needs regarding how the system responds to user input are catered to. Once the business requirements were gathered, a meeting was arranged with the stakeholders which consisted of multiple steps including group discussions, workshops, and questionnaires to gather the user requirements and finalize the design of ViDAS. The main idea behind conducting workshops was to observe the stakeholders while using a similar tool and document their approach as well as the steps taken in a data analysis task. With the help of these workshops, important user requirements were identified and gathered.

There were two workshops conducted and the purpose of both workshops was to gather important feedback on how the stakeholders want to perform the data analysis and visualization steps. The first workshop focused on Tableau, which is a business intelligence tool that allows its users to perform data analytics and create interactive visualizations to find patterns in the data and extract useful information. The second workshop consisted of Qlik Sense and focused more on data analytics including applying machine learning models on the data and later create interactive visualizations. The task lists designed for both of the workshops were fairly similar yet focused on their respective purpose and objectives. Crucial information was gathered during these workshops and important user requirements were pointed out during the discussions. After the workshops, the stakeholders were asked to fill questionnaires that are summarized in **Table 2**, giving brief feedback regarding the required user interaction and user experience the stakeholders expect.

With the help of the workshops and the questionnaires, the user requirements of ViDAS were collected. The user requirements focused on the drag-and-drop implementation of the tool during the data analytics and visualization processes. The custom analysis features of Qlik Sense and Tableau did not cover the basic requirements, due to which the custom analysis was prioritized to give ViDAS an advantage over these tools. The custom analysis should be created using a workflow pipeline and the resultant data should be exported to the visualization tab. The workflows in the custom analysis should comprise templates containing advanced machine learning models, and the ability to save and load custom workflows. Furthermore, the user should be able to write custom python scripts and execute them. The visualizations should be interactive and created using the charts as well as the fields. In the case of fields, ViDAS will automatically detect the dropped fields and create the best-suited chart based on those fields. The user requirements for ViDAS have been summarized below in **Table 3**.

Evaluation Summary		
	Qlik Sense	**Tableau**
Visualizations		
Data Overview	★★★★☆	★★★★☆
Chart-building Process	★★★★☆	★★★★★
Chart Recommendations	★★★★☆	★★★★☆
Interactive Visualizations	★★★★☆	★★★★☆
Custom Analysis		
Machine learning modules	★★☆☆☆	★☆☆☆☆
Visualize transformed data	★★★☆☆	★★★☆☆
Overall		
Overall Tool Execution	★★★★☆	★★★☆☆

Table 2.
An overview of Qlik sense and tableau evaluation (1 star - unforgettably bad, 2 stars - below average, 3 stars - average, 4 stars - above average, and 5 stars - unforgettably good).

ViDAS User Requirements
Interactive visualizations
Chart Recommendations
Drag-and-Drop design of the tool
Workflow pipeline for data analytics which is created using the drag-and-drop approach
Save & Load workflows
Pre-designed workflow templates
Custom python scripts for data analytics

Table 3.
Summary of ViDAS user requirements.

3.3 System requirements

System Requirements are the low-level requirements that act as the basic building blocks on which the system will be developed. These requirements cover all the technical details of the project including the technology used, the versioning, the compatibility, the database, and if the system will be hosted on a server. Once the user requirements were finalized, the next step was to define the system requirements based on the gathered user requirements. The discussions after the workshops gave a broad idea regarding the technology to be used in ViDAS development, thus shaping the system requirements as seen in **Figure 2**. It was required for the tool to be integrated into an already built website which was developed using React.js as a front-end and Java Spring as the back-end. Due to this limitation, ViDAS was

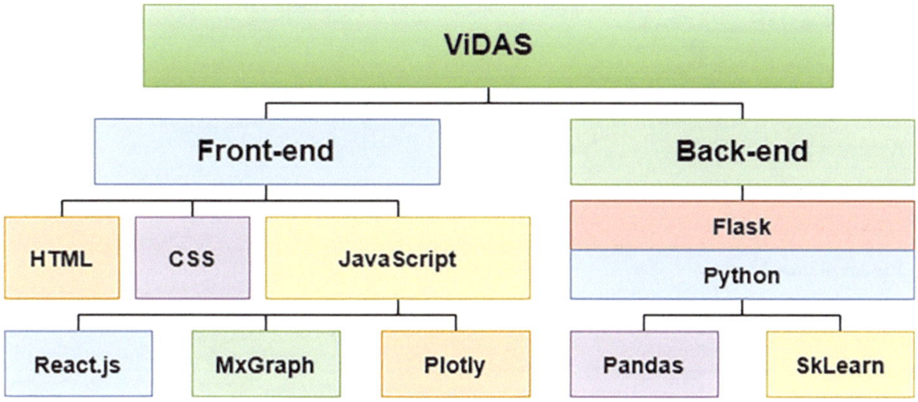

Figure 2.
ViDAS technology stack.

required to be developed using React.js as the front-end, while there was no restriction regarding the back-end implementation. After a detailed research, Python was decided as the back-end language due to it being highly dynamic for performing data analytics on big data and packaging a number of machine learning libraries. Flask was decided as the web framework for python to connect and communicate with the front-end.

For the front-end module, basic HTML was used to define the structure of ViDAS and CSS was used to style the HTML components for a complete user experience. React.js was used as the JavaScript framework to define the support of the functionalities in the front-end. To cater to the drag-and-drop requirement, MxGraph was used and customized according to the user requirements of ViDAS. The use of customized MxGraph further supported the User Interface and increased the intuitiveness of the tool while boosting the User Experience. Plotly was used to create interactive visualizations. These interactive visualizations were created in the back-end module and transmitted to the front-end for displaying them to the user.

The front-end was developed to run on the user's browser whereas the back-end was implemented to run on the server. The communication between the client and server was done using a promise-based HTTP API called Axios. With the help of Axios, the communication data was transmitted between the front-end and the back-end in JSON format. A state management library called Redux was used to manage the state and store the important data that was used by multiple pages on the front-end. Due to Redux, the data transmission among the front-end pages was eliminated and stored in a centralized state container.

The back-end is comprised of a number of Python libraries to deal with the big data and the data analytics functionalities. Pandas was used to store the big data in tabular form which helped in performing pre-processing steps before applying the machine learning models. Scikit-learn, a machine learning library, was used to support the machine learning models and provide various data analysis methods. The system requirements for ViDAS have been summarized below in **Table 4**.

3.4 Requirement outcomes

This subsection gives a brief outcome of this section and discusses the business, user, and system requirements that were gathered. The business requirements were gathered during the initial communication with the stakeholders and as a result, the scope of ViDAS and its users were identified.

ViDAS System Requirements
React.js as front-end language
mxGraph for drag-and-drop implementation
Redux for state handling
Python as back-end language
Flask as python web framework
Pandas to store big data
Scikit-learn for machine learning models
Plotly for data visualizations

Table 4.
Summary of ViDAS system requirements.

To gather the user requirements, two workshops were designed that comprised of task sheets. The stakeholders were guided through the tasks and their actions were observed. A number of new requirements were identified during this process. After the workshops, the stakeholders were asked to fill the open-ended and closed-ended questionnaires that gave shape to the user requirements of ViDAS. Once the user requirements were finalized, a number of post-workshop discussions were arranged where the technology stack of ViDAS development was discussed which gave shape to the system requirements of ViDAS.

All the requirements were gathered over a series of meetings, workshops, and discussions which have been summarized above in **Tables 1, 3**, and **4**. Now that the requirements of ViDAS were finalized, the next step in the human-centered design process was to produce the design workflows and to implement.

4. Producing design solution

The design creation process can be conducted in different ways depending on the scenario, from copying and development from previous design inspirations, to creating innovatively. Regardless of the source, all the design ideas progress through iterative development in the human-centered design approach. In such cases, mock-ups and simulations are essential to support this iterative cycle. Various design techniques are available such as brainstorming, parallel design, storyboarding, paper-based prototyping, and computer-based prototyping. While it is not intended to imply that all these techniques should be used in every product development, they should at least consist of a series of UI (User Interface) screens, and a partial database which allows the user to interact, visualize, and comment on the future design [23]. These early simulations are easy to create and result in a fault-free product at the end. The experts, stakeholders, and user representatives in the design development cycle help identify the faults, correct design, and undergo the costly process of re-implementation once the design is finalized.

After the requirements were finalized, as discussed in section 3, the next step was to create the design of ViDAS. Paper prototyping was used as a starting point to create the conceptual design of the tool. The stakeholders were involved in the design process to get their important feedback while creating the design. After the paper prototype feedback, the Balsamiq wireframing tool was used to transfer the conceptual design into interactive mock-ups [24] to better communicate the design with the stakeholders. These mockups consisted of design for each page of the tool and interactivity within the page (buttons, hyperlinks). Once the design was created and approved by the stakeholders, the development of ViDAS was started.

This section is divided into two subsections. The first one discusses the paper prototype of ViDAS. The next subsection addresses the software prototype and the implementation of ViDAS. These subsections will also address different tools that were used during this iterative process.

4.1 Paper prototyping

Paper prototyping is a widely used approach in the human-centered design process. It is a throwaway prototype technique used to create the initial conceptual design for a tool or application. Paper prototypes involve creating rough, even hand-sketched drawings or models of a design. The functionality is simulated by a member of the design team playing the computer and responding to the user's inputs by swapping the bits of the paper or writing an output. Creating paper prototyping is simple; however, it can provide convenient stakeholder feedback to aid the design [25].

Based on the requirements, paper prototypes of the different elements were drawn using pen and paper. These paper prototypes consist of data processing, chart creation workflow, and analytics workflow. Also, all the other necessary elements, such as menus, icons, buttons, labels, and dialog sequences were drawn. **Figure 3** shows the ViDAS paper prototype; each view has a description of what can be done and what happens when one interacts with individual elements.

The testing of the paper interface was video-taped as the elements moved and changed. This videotape of the paper prototype was shared with the stakeholders via email to get feedback about the initial design. Paper Prototyping was a handy technique during ViDAS initial design creation and getting the stakeholder's thoughts about the tool's overall shape. However, during the post-evaluation discussion of the paper prototype, it was found that the paper prototyping approach

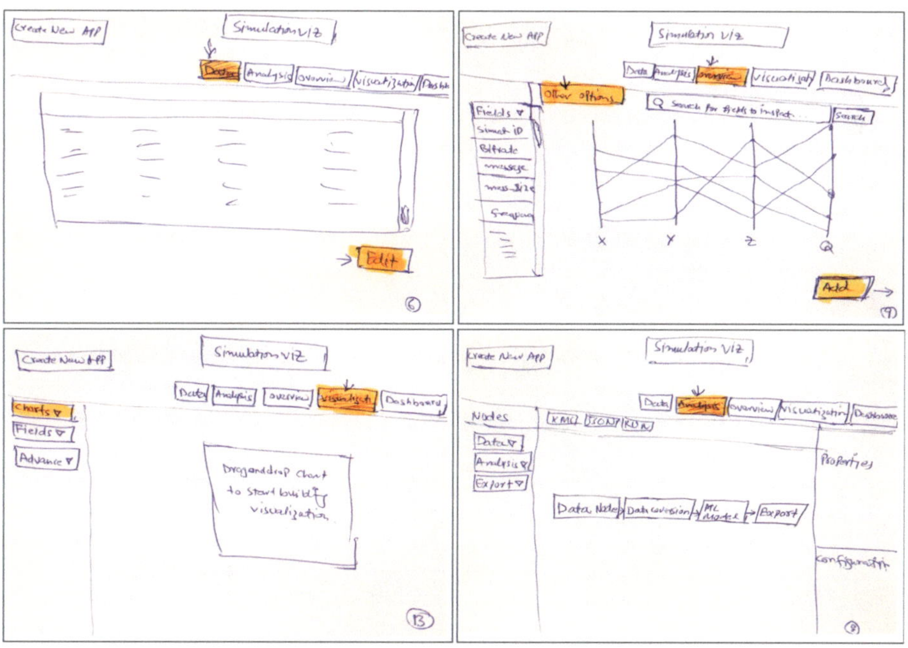

Figure 3.
ViDAS paper prototype.

does not discover every single usability problem. The paper prototype of the initial design was transformed into interactive mockups to address every aspect and give the design a more realistic feeling. It was also beneficial to communicate the design better with the stakeholders.

The mock-ups of ViDAS were created using Balsamiq [24] wireframing tool which is an industry-standard, light-weight wireframing tool used to create design mock ups and to show the interactivity among different pages and elements of the design. These mock-ups consist of each page of the tool along with interactivity that redirect when buttons or hyperlinks are clicked.

Figure 4 shows different design mock-ups of ViDAS. The Data tab view shows the tabular view of the uploaded file. Next to Data, the Overview tab is about visually inspecting the overall data. In contrast, the the Data Analysis tab shows the chart creation process that focuses on data fields. The Custom Analysis tab shows the workflow of creating analytics using network graphs. In this phase of the design, different changes were carried out which are summarized in **Table 5** and discussed later on.

ViDAS Mock Ups Evaluation.

Different changes were made to the design in this stage of the design creation process. The mockups creation process was iterative, and the focus group members were involved throughout this process. Mockups for different pages and interactivity among them were created, and the suggested changes were added iteratively until the final design was approved.

Table 5 shows all the stakeholders' suggested changes. These changes were added or deleted in multiple iterations in the ViDAS design. The label "Included" or "+" refers to components added to the design, while "Excluded" or "-"refers to the design components excluded from the final design. Some of these changes included

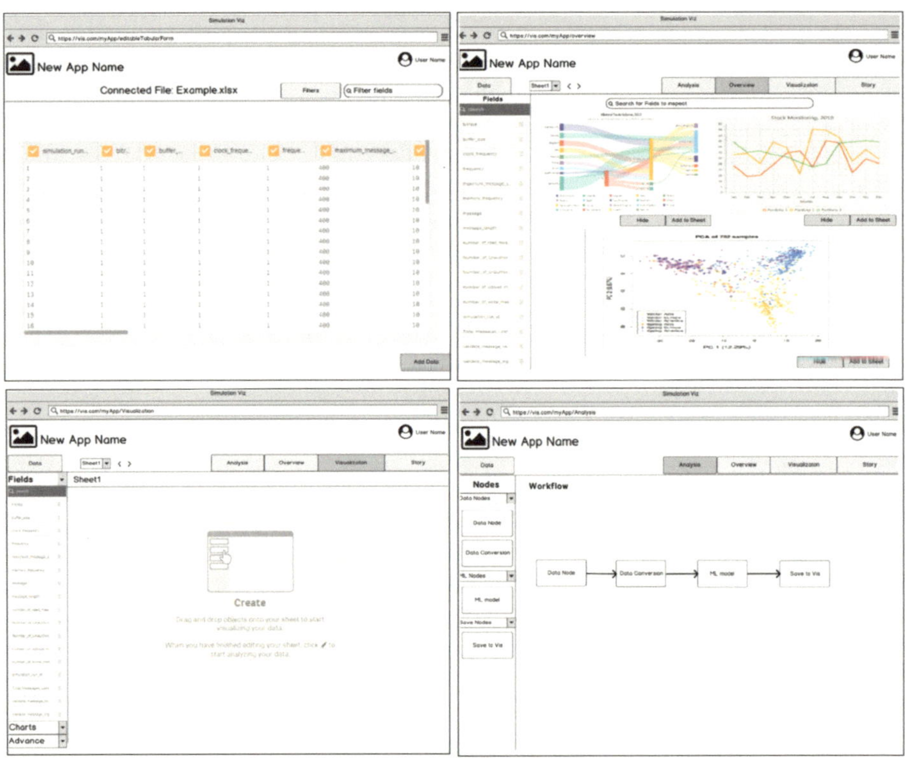

Figure 4.
ViDAS mock ups.

ViDAS Design Changes (– = Excluded), (+ = Included)			
Design Components	**First Iteration**	**Second Iteration**	**Final Design**
Login System	+	—	—
Color Combination	+	+	+
Button Labels	+	+	+
Tabs Concept	—	+	+
Sheet/New Sheet	—	+	+
Multiple Data Source Upload	+	—	—
Labels	+	+	+
Data Filtering	—	+	+
Data Tab	+	+	+
Overview Tab	+	+	+
Data Analysis Tab	+	+	+
Custom Analysis Tab	+	+	+
Story Tab	+	—	—
Dashboard Tab	—	+	+
Chart Per Sheet	—	+	+
Chart Creation (Drag and Drop)	—	+	+
Create Chart (Click and draw)	+	—	—
Create Chart (Selected Fields)	+	—	—
Drop-down menus	+	+	+
One Chart Per Sheet	—	+	+

Table 5.
Iterative ViDAS mock ups.

the User login system that was a part of the design but later excluded because ViDAS was planned to be an integrated module to a previous application and that application already has a user login system. Upload functionality was changed because currently the stakeholders use Excel data format for simulation output. Another design component was to create visualizations from selected fields exported from the Overview tab but was also excluded because the Overview Tab is currently supposed to only find trends and patterns in the uploaded data. Inspired by the tableau design, the concept of one chart per sheet (which would later be used for dashboard building) was also included in the design. Design for chart recommendation and sheets was also included later on. Further, changes such as labels of tabs, title page, and buttons were changed throughout this phase of the design.

4.2 Software prototyping

Once the requirements and design for ViDAS were finalized, the next step was to develop the tool or software prototype. ViDAS implementation covered all the requirements that were gathered in the requirement gathering process. The technologies that were used in ViDAS development allowed it to be integrated into the existing systems while still being easily extensible. ViDAS is implemented using the concept of RESTful APIs, where the front-end and back-end modules are separate and communicate with each other by sending the data over the network using their respective web addresses.

This section addresses the implementation of different components of ViDAS. It also discusses how and which component of the tool covered the requirements, such as applying analytics on raw data, visual overview of the uploaded data set, data preprocessing, and creating interactive visualizations.

The **front-end** of ViDAS is designed in such a way that when a user switches between different tabs, it keeps the previous page state intact. **Figure 5** shows different tabs of ViDAS. The **Data** tab is the entry point to ViDAS. When the file is uploaded, the tool shows the data in a tabular view initially. Thereafter the **Overview** tab gives a visual overview of the uploaded data. The **Data Analysis** tab handles the chart creation and chart recommendation process, and the **Custom Analysis** tab covers the data analytics process.

4.2.1 Data tab

Once a file is uploaded, the user is shown the uploaded file's tabular view. Data filtering can be carried out to delete unnecessary columns by clicking the **Edit** button at the bottom of the tab. If the data is clean and does not need filtering, the user can either switch to the Overview tab to see the data visually and find trends and patterns in the data, directly create charts by clicking the Data Analysis tab, or perform analytics on the data in Custom Analysis tab.

4.2.2 Overview tab

The overview tab covers the requirement "Visualize data to get an overview". This tab comprises three high-level charts of the uploaded data, as shown in **Figure 6**. The main objective of the Overview tab is to find trends and patterns in the data using a single interactive view.

The design of the Overview tab is divided into two sections. The left **side bar** consists of a drop-down for the high-level charts (Parallel-coordinates, Sankey

Figure 5.
ViDAS tabs.

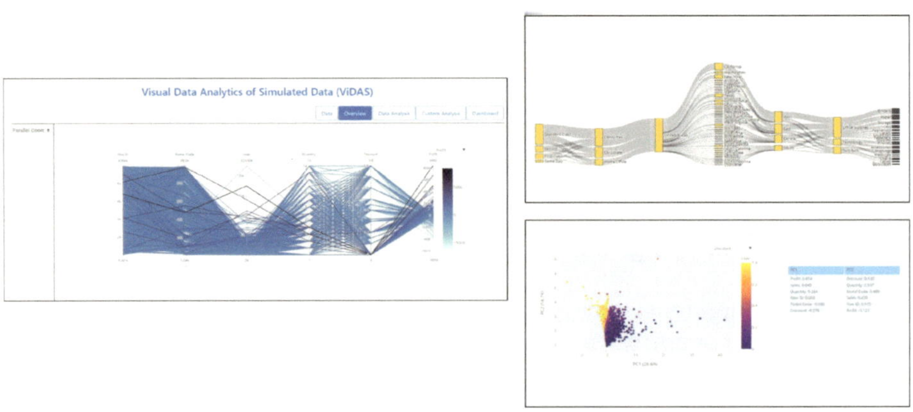

Figure 6.
ViDAS overview tab.

diagram, and PCA (Principal Component Analysis)), whereas the main canvas in the middle of the tab shows the selected type option from the drop-down. When the Overview tab is clicked, Parallel-coordinates is shown by default. Optionally, the user can switch to another chart option from the drop-down list. Here, Parallel-coordinates allows comparison of the individual observations (series) on a set of numeric variables. While Sankey diagram is a visualization technique that allows displaying data flow and PCA indicates variation and brings out strong patterns in a data set. It is often used to make the data easy to explore and visualize.

4.2.3 Data analysis tab

The Data Analysis tab addresses the requirements related to "Creating interactive visualizations" as stated by the stakeholders including chart creation by drag-and-drop, chart recommendations, and creating a chart from the exported custom fields from custom analysis tab.

The front-end of the Data Analysis tab is divided into different sections, as shown in **Figure 7** which consists of a sidebar, a toolbar, sheets on the bottom, and the recommendation charts drop down on the top right side. The **left side bar** comprises of all the drag-and-drop objects, such as standard and advanced charts, uploaded file's fields, and the custom fields exported from Custom Analysis tab. Similarly, the **top toolbar** consists of a text field which is used to rename or delete the active sheet and the **top right drop-down menu** is for chart recommendation functionality. The **Wrapper canvas** in the middle of the view allows the chart/fields to be dropped for chart creation. Furthermore, the **footer** holds all the sheets created by the user. The newly created data attributes which are exported from "Custom Analysis Tab" are shown in "Custom Fields" and can be used together with other fields for chart creation.

Charts in ViDAS are categorized into standard charts (e.g. line chart, bar chart, box plot, bubble chart, etc.), advanced charts (linear regression and correlation matrix), and overview tab charts (parallel coordinates, sankey chart, and PCA). Chart creation in the data analysis tab can be done in one of two ways. It starts when the user drops an object (chart/field(s)) to the canvas and provides the required parameters. The tool then generates the required chart for the user based on the provided data.

Create Chart By Chart Drop: Once a certain chart is dropped to the canvas, ViDAS first validates the type of the dropped object (chart/field(s)) and then

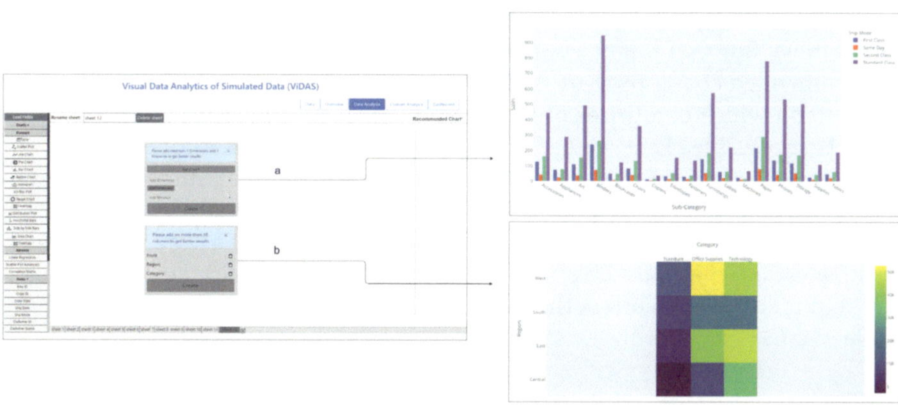

Figure 7.
ViDAS data analysis tab: (a) represents chart creation using chart drag-and-drop. (b) Represents chart creation using fields drag-and-drop.

renders a cart. The cart shows the dropped chart name on the top and a validation message about the required minimum parameters for the specific chart type for sensible chart creation. Once the required parameters are added and **create** button is clicked, the required chart is created.

Create Chart By Field(s) Drop: The chart creation by field(s) drop is similar to the chart drop. However, in the field(s) drop scenario, the user only drops the interesting field(s) as shown in **Figure 7**. ViDAS automatically calculates the type of the dropped field(s) as dimensions or measures. Thereafter, similar to the chart drop validation the user is shown a validation message with all the dropped field(s) for creating a proper chart. Here, the user may delete from the dropped field(s) as well. Once all the desired field(s) are dropped and **create** button is clicked, the tool creates the best possible chart by using the "Chart Recommendation" feature. At this point, the user can add additional fields or delete from the selected fields to update the created chart. The chart will change instantly and allows the user to see the changes visually while adding or removing certain fields.

Chart Recommendation: Once a chart is created, the user can switch to other possible charts in the "Recommended Charts" drop-down menu. This feature is known as 'Smart Charts' which was implemented in python and the resultant charts were created using the Plotly charting library. The feature collectively dubbed 'Smart Charts' is a collection of few chart recommendation methods. Once a chart is created, it suggests other possible charts that could potentially be made with the same fields that the user used to create the previous visualization so as to explore alternative charts that could better represent the data at hand.

The basic working of 'Smart Charts' depends on the attributes selected and their data types. Dimensions are the qualitative aspect of the data, in other words, the categorization and context of data whereas measures are the actual values of what you are measuring. Measures are always numeric and they are, most of the times, aggregated in some way using an aggregation function like sum, average, etc. The chart recommendation drop-down list contains all the charts for which the requirement of minimum dimensions and measures is fulfilled. The order in which they are displayed is decided by the weights that have been manually assigned to them. These weights are decided on basis of analysis of rival industry-standard data analysis tools, trial-and-error, and testing. A general pattern is, the more 'constrained' requirements that a chart has, the higher the weight it gets assigned. In other words, the more niche a chart is, the more important it would be if it can be created at all.

These recommendations in the "Recommended Charts" drop-down menu are listed in descending order of priority i.e. the higher up the entry, the better it is assumed to be. But that might not always be true, so it is up to the user to decide if the recommended charts present their data to their liking or not. 'Smart Charts' is a single method which comprises of various sub-modules including 'All possible charts', 'Best possible chart' and a 'Drag-and-Drop Chart Generator'.

If the user has an idea of what they want to create, they can directly choose a chart type, give it the required inputs and the specific chart would be generated. But if the user does not know what to create and they are roughly aware of the data fields in question, they can drag and drop the specific fields instead of the chart type. From there, ViDAS infers the dimensions and the measures to render the 'best possible chart'.

4.2.4 Custom analysis tab

The Custom Analysis Tab addresses the requirements related to the data analytics part of ViDAS that includes creating machine learning models for data analytics, pre-processing, and transformation of the data according to the user's requirements, and exporting the transformed data to the "Data Analysis Tab" for visualization.

Figure 8 shows the "Custom Analysis Tab", which consists of a **Side Bar** that contains all the drag and drop nodes. The user drags and drops these nodes onto the Canvas, which then creates a graphical representation of these nodes and can be connected later from their ports using edges. The **Properties Bar** shows the details of the selected Node, including the ID, Node Type, Node Name, the input, and output types. Once a node is configured, the user-defined configuration in the **Configuration Bar** is displayed. The **Tool Bar** includes the usability functionalities to help the user, including the option to save and load the workflows to and from the server, deleting the workflows, and creating a new workflow to clear the canvas.

The basic node design can be seen in **Figure 9**, with one input port and one output port. On top of the Node, there are details such as the node type and the custom name. The node type **Custom Node** is dynamic and changes according to the Node used. The node name **customName** can be changed by double-clicking on the Node. A square represents the input ports, and a triangle represents the output ports. At the bottom of the Node, there is an overlay that shows the current node state by using color indicators. When the user right-clicks on the Node, several options are available in the form of a drop-down menu. These options include Configure, Run, View Output, and Delete Node.

Initially, all the nodes are unconfigured when dropped, which can later be configured before being executed. ViDAS will not allow a node to be executed if that Node or any of its parent nodes are unconfigured. Some nodes require access to the data from their parent nodes to be configured, due to which they cannot be

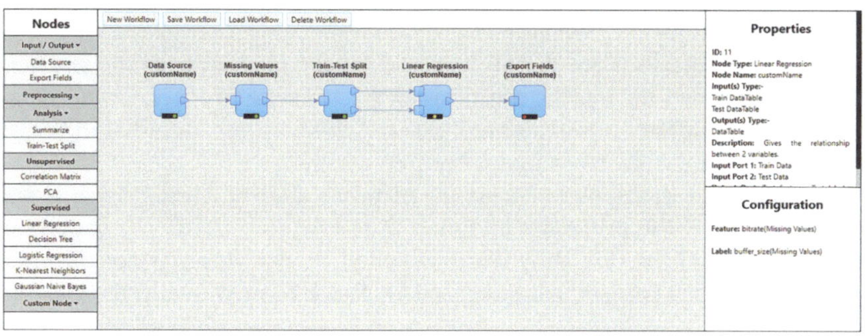

Figure 8.
ViDAS custom analysis tab.

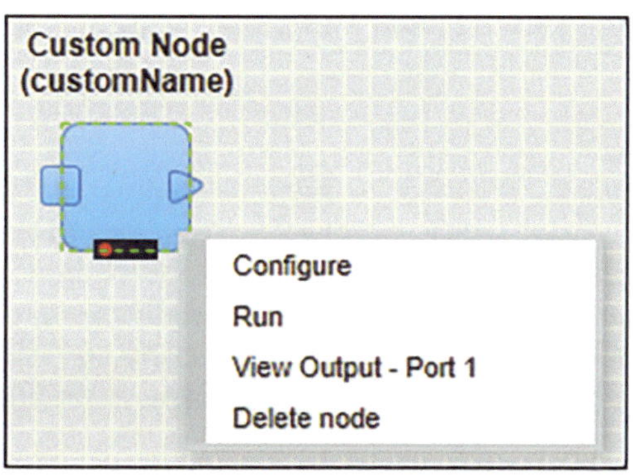

Figure 9.
Basic node.

configured unless the parent nodes are executed first. If multiple nodes have been configured in a workflow, the user can execute only the last Node, which will recursively execute all of the previous nodes up to and including that node.

ViDAS also allows the user to write custom python scripts to apply data analysis techniques that are not found in the built-in nodes. The user can select from a wide variety of node categories including input/output, preprocessing, analysis, and custom node to create the workflow pipeline.

Once the workflow pipeline is complete, the user has the option to export the transformed data to the "Data Analysis Tab" using the Export Node. The Export Node can be configured to select the required columns that the user wants to visualize. These exported columns can be loaded in the 'Custom Fields' drop-down list during data visualization. The feature to export the transformed data to another Tab gives ViDAS an integrated user experience which comprises of both data mining and interactive visualization methods.

5. Evaluating design solution

Evaluation is one of the most important components of project development and can play an important part in keeping the project aligned with its scope and objectives. Evaluation is used to support the decision making in development by applying evaluation methods and gathering the important data to compare the project to its pre-defined goals.

There are a number of evaluation types that exist but they can all be summarized into three most basic types which are Goal based, Process based, and Outcome based evaluations.

Goal based evaluations are mainly focused on the objectives of the project that have been pre-defined during the requirements gathering process. Once the project development is complete, it is evaluated to see if the features included in the project support the goals and objectives that were pre-defined.

Process based evaluations focus on the project's quality, strengths, and weaknesses. It discusses if the processes included in the project satisfy the stakeholders and are implemented the way they were intended. It also summarizes the strengths and weaknesses of the project as important feedback to improve in the next iterations.

Outcome based evaluations discuss the lasting effects of the project and the greater good that can be served as an outcome of the project. It measures the final goal of the project and how well it has been achieved.

This section will discuss ViDAS evaluation. It will also discuss the steps that were taken to evaluate ViDAS and to obtain important feedback, which was then used to improve the tool.

5.1 ViDAS evaluation

Once the implementation was done, the next step was to evaluate the ViDAS development. This evaluation was based on the business, user, and system requirements that were gathered during the requirements elicitation process discussed in section 3. First, the tool was self-evaluated and later by our stakeholders and users to get their useful feedback to improve the User Experience and User Interface of ViDAS.

The evaluation of ViDAS was mainly goal-based as the development was done keeping in mind the requirements and objectives that were defined before the implementation. Various evaluation methods were combined for ViDAS as shown

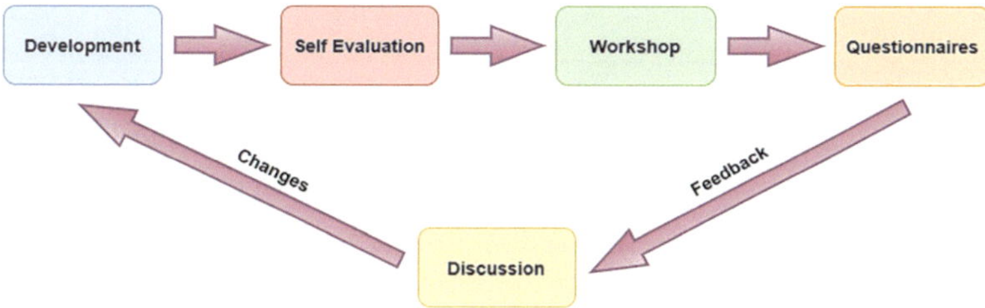

Figure 10.
ViDAS evaluation steps.

in **Figure 10**, including comparing the goals to the pre-defined documents, arranging a workshop with the focus group, and collecting important data before and after the workshop. The evaluation also partially included segments from process based and outcome based evaluations as the project's quality was assessed, important feedback was collected, and the final goal was studied during the complete process.

Once the development of ViDAS was completed, the features and functions developed were compared to the pre-defined requirements. The product was assessed in the view of the objectives and goals defined during the requirements gathering process. Also, a number of tests were carried out that included analyzing the reliability of the front-end on multiple browsing agents as well as various screen sizes. Stress tests were also a vital part of the self-evaluation process and ViDAS was tested by uploading big data containing millions of rows to assess the data handling time and look for any bottlenecks that may hinder the tool's performance. All of these steps were an important part of self-evaluation and the necessary changes were made to improve the final User Experience (UX) before preparing a workshop with the stakeholders.

After self-evaluation, a workshop was arranged with the stakeholders. This workshop was prepared using tasks similar to the ones used during the requirements gathering process to compare and improve with the state-of-the-art in the market. Similar tasks made it easy for the stakeholders to compare ViDAS with the tools used in the requirements gathering workshop.

5.2 Feedback

After the evaluation workshop, the initial feedback was collected from the stakeholders and documented to improve ViDAS. The stakeholders were pleased with the overall implementation of the tool. While there were suggestions to improve ViDAS, it was recognized to be a great help with the analysis of the simulation data for the stakeholders.

The post-workshop feedback was gathered by asking the stakeholders to fill open-ended and closed-ended questionnaires that have been summarized in **Table 6**. Overall, the stakeholders were satisfied with the tool execution. The data overview, the chart building process, and the interactive visualizations were greatly appreciated. The stakeholders also found the chart recommendation concept useful. The custom analysis features were commended and the workflow pipeline design which comprised of the drag and drop module was praised by the stakeholders. Furthermore, the development of ViDAS in such a short span of a few months was greatly appreciated, which resulted in the overall feedback being much positive. However, the overall tool execution feedback was lower than anticipated and this was mainly due to some usability aspects such as missing validations, missing tool tips, and labels not being user-friendly. These usability issues have since been resolved and have been implemented in the second iteration of the tool.

ViDAS Evaluation Summary	
Visualizations	
Data Overview	★★★★☆
Chart-building Process	★★★★★
Chart Recommendations	★★★☆☆
Interactive Visualizations	★★★★☆
Custom Analysis	
Workflow Pipelines	★★★★★
ML modules without coding	★★★★★
Export Data for Visualizations	★★★★★
Overall	
Overall Tool Execution	★★★☆☆

Table 6.
ViDAS evaluation feedback overview (1 star - unforgettably bad, 2 stars - below average, 3 stars - average, 4 stars - above average, and 5 stars - unforgettably good).

6. Conclusion and future work

6.1 Conclusion

The demand for a good visual analytics tool is increasing due to the need to cater to complex data which is constantly increasing. A complete visual analytics solution should cover all the procedures in the conventional visual analytics pipeline. These procedures, starting from the basic data pre-processing to clean the data, allow the user to apply machine learning methods in order to transform the data, and extract hidden correlations in the data. Later, the transformed data can be visualized to find trends and patterns that can be crucial in extracting knowledge from the data.

While there are various tools available in the market that allow the users to either create dynamic, interactive visualizations or implement machine learning models, none of the tools cover the complete visual analytics pipeline with their built-in features. These tools rely on third-party extensions that can be quite tiresome to configure and integrate. Additionally, integrating these third-party extensions may make the tools quite heavy and unstable. This project discusses the visual analytics pipeline and presents a solution that implements the complete visual analytics pipeline in order to fill this void. In this project, a web-based tool called ViDAS was developed using the "Human-centered approach" that packages the complete visual analytics pipeline integrated into it. ViDAS was developed with a number of requirements in mind that are not available in the tools currently in the market. These requirements included the tool being light-weight so that it can be deployed on a web server, dealing with big data, allowing machine learning methods and custom scripts, and visualizing the data in an interactive way.

Initially, the context of use for the tool and requirements were gathered in a series of workshops and meetings. ViDAS allows it's users to upload raw data and apply initial data filtering to only process the required attributes. These attributes can later be used to either create interactive visualizations or apply custom analysis.

If the custom analysis is applied, the resultant attributes can be exported to the 'Data Analysis' tab for visualization of those ML models. The chart creation process of ViDAS comprises of a drag-and-drop workflow and is very intuitive. Furthermore, ViDAS packages a chart recommendation functionality that suggests a number of charts according to the fields selected. By comparing the feedback of the tools used in the requirement gathering process to the feedback of ViDAS in its evaluation, an improvement was seen in the overall tool experience.

6.2 Future work

In this project, a complete visual analytics tool was presented that can help the users in finding hidden relationships in their data by providing various machine learning and visualization techniques. However, the tool is still in its initial phase and requires more features in order to cater to all the needs of a data analyst.

In the initial ViDAS implementation, there was an intention to include dashboards, which would feature the coordinated views, cross-filtering i.e. filtering/selecting data in one chart would reflect relevant changes in other linked charts. Dash, an extension to the Plotly charting library, was explored to create these coordinated views but unfortunately, this could not be completed due to time constraints and may be added in the future.

Another interesting feature that could be implemented in the future is the functionality to select desired fields in the overview tab. During the overview tab view, the user may find multiple correlated fields that are important to visualize. These fields can be intuitively selected from the overview visualization to be used in the desired visualizations. Furthermore, an area that can be significantly improved further is the processing of date/time data during the visualizations. Currently, dates are treated as any other dimension or ignored. There could possibly be sophisticated parsing by using parts of dates such as years and months. Also, interactivity within the visualizations can be further improved as Plotly features a handful of interaction widgets that can be added.

The functionality of 'Smart Analytics' can be integrated into the tool to help the user in applying the machine learning models. In some cases, the raw data can be unstructured and complex which may pose a challenge in deciding which machine learning models to apply. In such cases, the smart analytics feature can use data mining techniques to explore the type of data in the data set and suggest the recommended machine learning models to apply based on the data type and complexity. Additionally, more ML nodes can be introduced that offer flexibility in dealing with complex data. Forecasting can be used with time-series data to predict future values depending on the past data-set.

Furthermore, due to the current situation of a pandemic, the evaluation of ViDAS could not be performed to the full extent. The evaluation was conducted by using only the stakeholders as the focus group. In the future, a complete evaluation should be conducted that includes a mixed focus group comprising of the domain experts as well as the non-technical users.

These new features combined with the in-depth evaluation of ViDAS can result in a better and thorough visual analytics tool that provides a user-friendly experience and helps its users in dealing with raw and complex data by allowing data analytics as well as interactive visualizations.

Acknowledgements

The authors wish to thank the members of the Human Computer Interaction Lab at the University of Kaiserslautern and the Embedded Software Engineers at the

Fraunhofer IESE for their cooperation. This work was supported by the Ministry for Science, Education, and Culture Rhineland-Palatinate through the Virtual Engineering of Smart Embedded Systems (ViSE) project.

Author details

Taimur Khan*†, Syed Samad Shakeel†, Afzal Gul†, Hamza Masud and Achim Ebert†
University of Kaiserslautern, Kaiserslautern, Germany

*Address all correspondence to: khan@rhrk.uni-kl.de

† These authors contributed equally.

References

[1] Chen M, Mao S, and Liu Y. Big data: A survey. Mobile networks and applications 2014; 19:171–209

[2] Lee JY, Kim JY, You SJ, Kim YS, Koo HY, Kim JH, Kim S, Park JH, Han JS, Kil S, et al. Development and Usability of a Life-Logging Behavior Monitoring Application for Obese Patients. Journal of Obesity & Metabolic Syndrome 2019; 28:194

[3] Smith KT. Needs analysis: Or, how do you capture, represent, and validate user requirements in a formal manner/notation before design. Human Factors and Ergonomics in Consumer Product Design: Methods and Techniques 2011; 415

[4] Nielsen J. Usability 101: Introduction to usability. 2012. Available from: https://www.nngroup.com/articles/usability-101-introduction-to-usability/

[5] Kleijnen JPC, Sanchez SM, Lucas TW, and Cioppa TM. State-of-the-Art Review: A User's Guide to the Brave New World of Designing Simulation Experiments. INFORMS J. on Computing 2005 Jul; 17:263–89. DOI: 10.1287/ijoc.1050.0136. Available from: https://doi.org/10.1287/ijoc.1050.0136

[6] Kehrer J and Hauser H. Visualization and Visual Analysis of Multifaceted Scientific Data: A Survey. IEEE Transactions on Visualization and Computer Graphics 2013 Mar; 19:495–513. DOI: 10.1109/TVCG.2012.110

[7] Post T, Ilsen R, Hamann B, and Aurich J. User-Guided Visual Analysis of Cyber-Physical Production Systems. Journal of Computing and Information Science in Engineering 2016 Oct; 17. DOI: 10.1115/1.4034872

[8] Tableau. Business Intelligence and Analytics Software. https://www.tableau.com/. [Online; accessed 25-February-2020]

[9] Intelligence QB. Data Analytics and Data Integration. https://www.qlik.com/us/. [Online; accessed 25-February-2020]

[10] Power BI - Find clarity when you need it most. https: //powerbi.microsoft.com/enus/. Accessed: 2020-09-24

[11] KNIME - End to End Data Science. https://www.knime.com/. Accessed: 2020-09-19

[12] RapidMiner Webpage. https://rapidminer.com/. Accessed: 2020-09-20

[13] Orange - Data Mining Fruitful and Fun. https://orange.biolab.si/. Accessed: 2020-09-19

[14] WEKA Webpage. https://www.cs.waikato.ac.nz/ml/weka/. Accessed: 2020-09-20

[15] KEEL - Knowledge Extraction based on Evolutionary Learning. http://www.keel.es/. Accessed: 2020-09-19

[16] Wang XM, Zhang TY, Ma YX, Xia J, and Chen W. A Survey of Visual Analytic Pipelines. Journal of Computer Science and Technology 2016 Jul; 31: 787–804. DOI: 10.1007/s11390-016-1663-1

[17] Keim DA. Information Visualization and Visual Data Mining. IEEE Transactions on Visualization and Computer Graphics 2002 Jan; 8:1–8. DOI: 10.1109/2945.981847. Available from: https: //doi.org/10.1109/2945.981847

[18] Altalhi AH, Luna JM, Vallejo M, and Ventura S. Evaluation and comparison of open source software suites for data mining and knowledge discovery. Wiley Interdisciplinary Reviews: Data Mining and Knowledge Discovery 2017; 7:e1204

[19] Kotu V and Deshpande B. Predictive analytics and data mining: concepts and practice with rapidminer. Morgan Kaufmann, 2014

[20] Young RR. Recommended requirements gathering practices. CrossTalk 2002; 15:9–12

[21] Requirements Gathering: Types and Methods. https://digitalmarketing.temple.edu/agervasio/2017/07/18/requirements-gathering-typesand-methods/. Accessed: 2020-11-11

[22] Kazhamiakin R, Pistore M, and Roveri M. A framework for integrating business processes and business requirements. *Proceedings. Eighth IEEE International Enterprise Distributed Object Computing Conference, 2004. EDOC 2004*. IEEE. 2004 :9–20

[23] Maguire M. Methods to support human-centred design. International Journal Human-Computer Studies 2001; 55:587–634

[24] Faranello S. Balsamiq wireframes quickstart guide. Packt Publishing Ltd, 2012

[25] Snyder C. Paper prototyping: The fast and easy way to design and refine user interfaces. Morgan Kaufmann, 2003

Chapter 8

Introduction to Intelligent User Interfaces (IUIs)

Nauman Jalil

Abstract

This chapter is intended to provide an overview of the Intelligent User Interfaces subject. The outline includes the basic concepts and terminology, a review of current technologies and recent developments in the field, common architectures used for the design of IUI systems, and finally the IUI applications. Intelligent user interfaces (IUIs) are attempting to address human-computer connection issues by offering innovative communication approaches and by listening to the user. Virtual reality is also an emerging IUI area that can be the popular interface of the future by integrating the technology into the environment so that at the same time it can be more real and invisible. The ultimate computer interface is more like interacting with the computer in a dialog, an interactive environment of virtual reality in which you can communicate. This chapter also explores a methodology for the design of situation-aware frameworks for the user interface that utilizes user and context inputs to provide details customized to the activities of the user in particular circumstances. In order to comply to the new situation, the user interface will reconfigure itself automatically. Adjusting the user interface to the actual situation and providing a reusable list of tasks in a given situation decreases operator memory loads. The challenge of pulling together the details needed by situation-aware decision support systems in a way that minimizes cognitive workload is not addressed by current user interface design.

Keywords: Intelligent User Interfaces (IUIs), Human-Computer Interaction (HCI), Graphical User Interfaces (GUI), Ubiquitous Computing (Ubicomp), User-Centered Design (UCD)

1. Introduction

The methods by which people have interacted with computers have made great strides. The journey continues and new designs of systems and technologies emerge more each day, and in the last few decades, research in this field has expanded quite rapidly. Artificial intelligence tries to simulate human abilities with the help of machines. This challenge, launched at the famous Dartmouth conference in 1956, has attracted a great deal of interest and effort from the research community over the last 50 years. While the community has not been able to accomplish the euphoric expectations presented at that conference even after so many years, a good number of accomplishments were achieved in this simulation of human capabilities with computers, embracing the progression of the artificial intelligence field.

One of the cornerstones of simulating human abilities, without any doubt, is communication. Communication between humans, between machines, or between humans and machines can be realized in this scenario. In communication theory, communication between humans has indeed been studied for several years, communication between machines is still currently a major topic in conferences, with a peculiar focus on ontologies and communication acts. Finally, in the Human-Computer Interaction discipline, communication between humans and machines is being studied. The design, implementation and assessment of user interfaces are discussed in this discipline.

From the first command-based user interfaces to the most advanced graphical ones, the brief, but extremely active, history of computer science has witnessed a real revolution in user interfaces. The computational balance is now leaning towards interaction, cultivating an increasing interest in user interfaces. At the same time, the growing power and sophistication of user interfaces is encouraging the creation of techniques and modalities of interaction closer to human beings' cognitive models. In the pursuit of these goals, several different artificial intelligence methods have been incorporated into user interfaces over the last fifty years to provide a more natural and efficient interaction between humans and computers.

One of the key parts of an application is the user interface. If it is designed correctly, when communicating with the computer, the user may feel relaxed. On the other hand, if an application is capable of performing the tasks for which it was intended, the application would still not be acceptable to users unless it is capable of interacting in an intelligible and functional way with the user. There is a real interest in enhancing communication between the user and the machine within the Human-Computer Interaction group, recognized by the large number of researchers dedicated to the study of techniques intended to improve usability [1] in a user interface.

The Human Computer Interaction research group is looking for how to enhance the user's feeling from the user interface in the quest to build user interfaces with a high degree of usability. One main problem to be tackled in this pursuit is to make the interaction even more natural. This growing interest in making interaction even more natural means raising the standard of communication between humans and machines. The fundamental concepts of natural interaction have led to the exploration and implementation of a large number of techniques in interaction design coming from Artificial Intelligence, enabling, among many other things, more detailed responses to the actions of users. The term Intelligent User Interfaces have been created by this integration of Artificial Intelligence techniques within Human-Interaction Techniques, where this work is immersed.

The framework for the design of situation-aware interfaces are also covered in this chapter, in such a way that input information (context and environmental indications) can be specifically taken into consideration in the task specification [2]. It is believed that the designer adds abstract UI components to the task model to improve a concrete user interface (UI). This information is platform-independent, so that this information can ultimately be used by the rendering back-end to build a specific UI for different platforms. The next phase includes developing the model for dialogue. In order to simplify the work of designers, designers can be helped by automatically creating the state and transformations between the different individual dialogues. The method provides an algorithm to measure the various dialogues and transformations from the mission definition between dialogues. These transformations can be modified, incorporated or omitted by designers according to the outcome of a previous testing stage or the expertise of the designers. This allows situation-aware UI programmers to exploit transformations caused by changes in the situation. Therefore, programmers have power over the effect of the condition on the UIs' usability. In [3] research focuses on developing experiences of visual context-aware services by integrating approaches from computer vision and artificial intelligence. It has close connections to the field of intelligent user interfaces.

The touchscreen keyboard is the most prevalent intelligent user interface on modern cell phones, and it is vital for mobile communication. Working to develop smarter, more effective, easy-to-learn, and enjoyable-to-use keyboards has raised a slew of intriguing IUI interface and research questions [4]. The progress and open research questions over the last decade in text input, emphasis on and directly dealt with through publications, including robotics and estimation cost-benefit equations [5], the significance of human performance models in the creation of error-correction algorithms and the potential of machine/statistical intelligence [6–11], the ramifications of spatial scaling from a phone to a watch on human-machine labour separation, consumer behaviour and learning creativity, and the complexities of assessing the longitudinal impact of personalization and adaptation are discussed in [12–17]. The aim of this study is to show that intelligent user interfaces, or the integration of artificial intelligence and human factors, are the future of human-computer interaction and information technology in general.

By varying decision factors' forms, numbers, and values, this article [18] provides a mechanism for adaptive, measurable decision making for Multiple Attribute Decision Making (MADM). This research can be used to help designers create intelligent user interfaces for HCI decision-making applications that respond to user experience and decision-making efficiency. In [19], a Genetic Programming-based technology is proposed for automating crucial design phases. Designers can specify simple content elements and ways to merge them in this method, which will then be automatically composed and checked with actual users by a genetic algorithm to find optimal compositions.

2. Human computer interaction

The idea of Human-Computer Interaction/Interfacing (HCI) was automatically represented with the advent of the computer or, more generally, the system itself, also referred to as Man-Machine Interaction or Interfacing. In fact, the explanation is straightforward: most advanced machines are useless unless men can use them properly. The main ideas that should be considered in the development of HCI clearly present in this simple argument: functionality and usability [20].

In the end, why a system is actually built can be determined by what the system can really do, i.e. how a system's function can assist in achieving the system's purpose. A system's functionality is characterized by the collection of actions or services it offers its users with. After all, the functionality value is only observable until it becomes feasible for the user to be used effectively [21]. The usability of a system with any functionality is the extent and degree to which the system can be used easily and reasonably by certain users to achieve specific objectives. If there is an appropriate equilibrium between the functionality and usability of a system, the real productivity of a system is achieved [22].

Taking these definitions into account and knowing that in this context, the term system, machine and computer are often used interchangeably, HCI is a design that should build a match between the user, the machine and the services needed in order to achieve a certain output both in terms of quality of service and optimality [23]. Most of it is subjective and context based to decide what makes a certain HCI design successful. An aircraft model design tool, for instance, must have precision in the display and design of the components, whereas graphics editing software doesn't really require such accuracy. The technologies available may also have an effect on how various types of HCI are configured for the same reason. For example, commands, menus, graphical user interfaces (GUIs) or virtual reality can be used to access the information on a given device. A more detailed description of current techniques and technologies used to communicate with computers and the latest developments in the field is provided in the next section.

3. Intelligent user interfaces

Intelligent user interfaces are human-machine interfaces whose purpose is to enhance the performance, affectivity, naturalness and, in general, usability of interactions between humans and machines representing, reasoning or performing on a collection of models (user, domain, dialog, voice, functions, etc.). The design of user interfaces is a multidisciplinary challenge due to the numerous models coming from different disciplines (see **Figure 1**). Artificial Intelligence leads to developing collaboration with intelligence modeling methods, Software Engineering corresponds to coherent systems, notations and formal languages. Consequently, the user's concern leads to Human-Computer Interaction, and hence the strategies for designing practical user interfaces.

In order to fulfill the key purpose of intelligent user interfaces and to help them in various situations, they need to be able to reflect the information they have about users, the activities they are allowed to do across the user interface, the usage sense in which the user communicates with the program and has the ability to properly interpret the inputs and produce the outputs based on all the data gathered and the information they have [25]. The topic of usage is generally defined by presenting a model of the application users' strengths, skills and interests, the channels on which those users communicate with the application (both the hardware and the software platform), and the physical world in which the interaction takes place, such as luminosity, noise level, etc. In [26], the task actually performed by the client is often included as a first order element in the scope of use. Since the current role is included in the framework of use, given a particular context of use, it is possible to better explain the reactions the machine would give. It can, for example, be used to build context-sensitive support systems quickly. It is very helpful to include the task in the context of use. It is not always feasible, though, to provide a high level of assurance about what the individual is actually carrying out. According to the interaction data obtained from the client, the authors add certain heuristics to a task model to figure out which is the task that is actually being done. The heuristics work

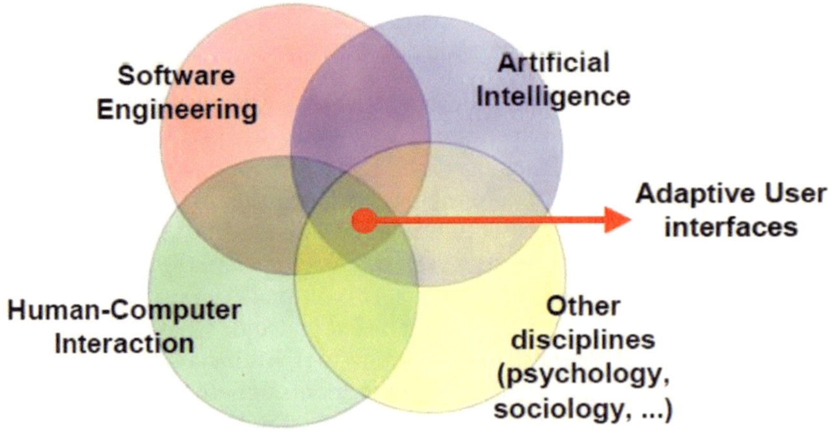

Figure 1.
Various disciplines in the conception of intelligent user interfaces [24].

on the premise that the possible tasks that the user executes at a given time are just the set of tasks "enabled" in the actual presentation in which the user communicates at all times [27].

3.1 Intelligent user interfaces goals

In **Figure 2** a few of the most critical issues encountered in intelligent user interfaces are illustrated [28]. These challenges are targeted at achieving the ultimate target of intelligent user interfaces: improving the overall usability of the machine.

As in **Figure 2** the reader will observe that certain human capacities are included, such as understanding, adaptation, logic or simulation of the world. Many of those strengths are study lines established from the very beginning of this field of Artificial Intelligence. Many academic reports on artificial intelligence claims that intelligence is a program that can be executed regardless of the platform on which it is executed, irrespective of whether it is a machine or a brain. Two models, the user conceptual and the execution of one of the machines, must create a sort of equilibrium and comprehension when the user communicates with an item. The user uses a concrete vocabulary of operation defined by the input devices and the metaphor for which the user interface has been developed. On the other side, by conforming the contents, displaying it, the artifact must interpret the feedback and respond to the interaction, and it should also be able to determine the interaction mechanism itself and make conclusions as to how useful it is.

There are several areas of Artificial Intelligence that bring approaches, strategies and ideas of vital importance to the development of intelligent user interfaces. Methods for providing learning capacities such as neural networks or Bayesian networks can be used in Artificial Intelligence, knowledge representation methods such as semantic networks or frames, decision models such as fuzzy logic, expert systems, case-based reasoning and decision trees, etc. We may work on the user interface by applying these inputs to the intelligent user interface by facilitating transformation, making its use simpler, analyzing interaction, simulating activities, leading the user or assisting the developer of the user interface.

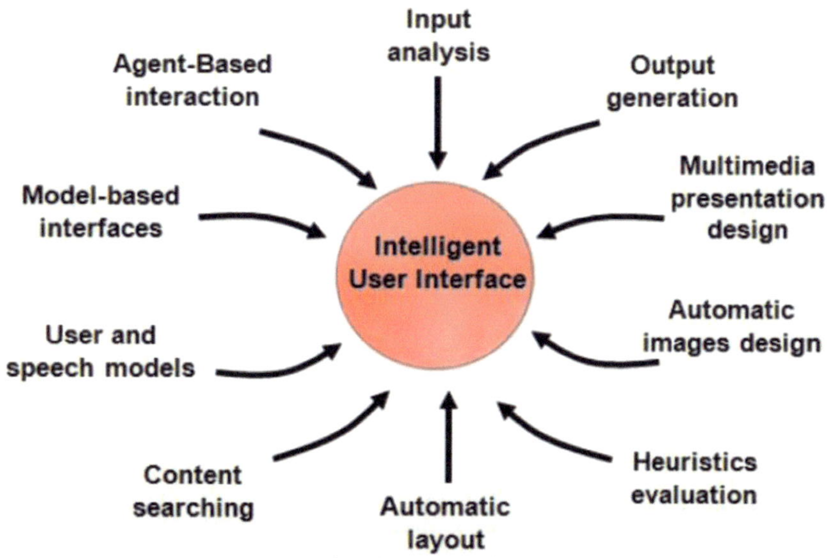

Figure 2.
The most common objectives seen in intelligent user interfaces [24].

4. Ubiquitous computing and ambient intelligence

The most recent HCI study is unquestionably ubiquitous computing (Ubicomp). The phrase frequently used by ambient intelligence and pervasive computation interchangeably, refers to the ultimate human-computer interface techniques, which are the elimination of a desktop and the integrating of the computer in the environment in order to make it unseen to humans when covering them everywhere. Mark Weiser first proposed the concept of ubiquitous computing during his time as chief technologist at Xerox PARC's Computer Science Lab in 1998. His vision was to incorporate machines into the world and ordinary objects anywhere so that individuals could connect with several computers at the same time because they are hidden to them and communicate with each other wirelessly [29].

The Third Era of computation has also been called Ubicomp. The first era was the mainframe era, with one machine and several people. Then it was the Second Era, one user, one computer, named the PC era, and now Ubicomp introduces more machines, one person era [29]. The major developments in computation are seen in **Figure 3**.

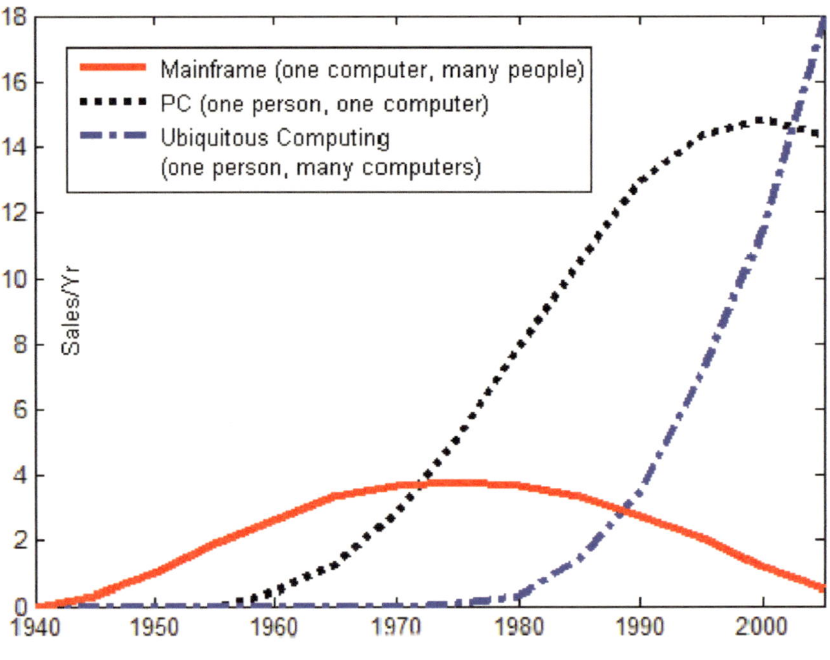

Figure 3.
Major trends in computing [29].

5. Intelligent user interface development

The predominant trend since some years ago is model-based user interface development (MB-UID) in software product engineering and user interface is not an exception [30, 31]. The key concept behind this pattern is to define a set of templates representing the characteristics of the user, the activities, the domain, the sense of usage, and the user interface at various levels of abstraction in a declarative way. A model is a generalized representation of a portion of the world called the system. All of such models are typically stored in the description language of XML-based user interfaces. For intelligent user interfaces, XML has become a de facto *lingua franca* [32]. Following this strategy, intelligent user interfaces [33] and hypermedia systems [34] are now being developed today.

Knowledge bases store the information that the device has on both the program being performed and the sense of use. During the various stages of design, the information concerning the program is often collected. In our case, the meta-model used to store information is a slightly updated version of the usiXML1 definition language of the XML-based user interfaces. While description languages like XIML2 or UIML3 may be used for other user interfaces, too. The user interface information involves:

- The *domain model* has to hold certain objects/data the user interface requires for the user to conduct his activities.

- The *task model* that represents the activities to be done by the user using the user interface and the temporal limits between those activities. A notation based on ConcurTaskTrees is used to model it [35].

- The *abstract user interface model* comprises the user interface represented in terms of abstract interaction artifacts [36], and irrespective of the platform (it does not focus on the functionality of various types of systems, such as PCs, PDAs or mobile phones) and regardless of the modality (graphical or vocal). This model is very helpful for connecting the spaces of interaction for which widgets are eventually positioned with the objects/data used to execute the tasks in those spaces of interaction [37].

- The *concrete user interface model* depicts the user interface, but in this scenario it is comprised of actual objects of interaction [36]. In this context, the representation of the user interface is based on the platform, and it will be the primary model used to create a complete user interface with which the user will eventually communicate.

- The *context model* keep all the details that the device will gather or handle about the sense of usage. It contains a user model, a model of its platform, and a model of its environment.

The user model contains certain user features important to the framework (preferences, skills, knowledge). Applying user-modeling techniques [38, 39] to the input data obtained by the sensors and the information already processed, this model is updated. For example, all the activities done by the client in the user interface are documented in the interaction log, so that the machine can use data mining or classification techniques to determine new user information to update the user model.

The platform model comprises the functionality (both hardware and software) of prospective system profiles where the program can be executed. By analyzing the input data from sensors, this model is also updated. For example, if the user increases the screen resolution, the visualization space available for viewing the contents is limited, and so the document structure, or even the contents themselves, is likely to change. Therefore, as the user modifies the screen resolution, the adjustment is identified by a device sensor so that the platform model can be changed accordingly.

The environment model contains the information about the physical world in which contact is performed. The potential amount of information collected from the surroundings would obviously be extremely large, and the developer must therefore determine which information is important because it has an effect on the application's use. Once more, the data stored in this model is regularly updated by the incoming sensor data. Good relationship management means that all the information contained in the knowledge bases of the previously mentioned models is used efficiently and effectively.

The human-computer interaction utilizes the nature of intelligent user interfaces, leading to the problems that a user interface should still have about access and quality of use [40]. *Usability describes the degree to which individual consumers may use a product to accomplish given objectives with performance, effectiveness and satisfaction in a specified sense of use (ISO 9241-11).* The metaphors already mentioned, the increasing computing capacity and the modern ways of communicating are encouraging the emergence of new methodologies that encourage the consideration of artifacts of interaction other than those historically considered [41]. User-centered programming requires or at least draws influence from each of these methodologies (UCD). User-Centered Design (UCD) is a strongly organized, systematic approach for product creation driven by: (1) specifically identified market priorities that are task-oriented and (2) awareness of user expectations, constraints and desires.

6. Designing situation-aware decision support systems

An outline of the design process is given in this section (**Figure 4**). Describing the situation-aware user interface, the design process encourages the layout of declarative abstract models [2].

To export these models to the runtime, the aggregate of the models can be serialized. The corresponding UI can be created in the form of a prototype to check the responsiveness of the device in order to evaluate the outcome of these models. Considering the prototype, certain adjustments may be made to the templates in the modeling process to improve the appearance of the UI, for example, or how changes in the scenario can affect the UI.

Situation-based Task Model: Firstly, a work model is defined that specifies the tasks that users and applications can experience as they communicate with the system. Since we want situation-aware UIs to be created, tasks often focus on the current situation. This is why activities are drawn up for different circumstances in the task model. The designer will specify various roles for different scenarios in this way.

Input Model: The developer needs to identify what kind of input will affect the interaction, i.e. the tasks, when the task model is defined. This can be achieved by choosing the input collection objects (Perception Objects or POs). The aggregation objects (AO) may be aggregated by these objects and described by interpretation objects (IO). The designer will do this by binding AOs to POs and choosing how the information needs to be interpreted from a set of predefined interpretation rules. At the comprehension layer, the IOs reflect the interpreted data. The designer has to connect the IOs to task model nodes when the input model is defined (intermodal connection). The designer may thus denote which activities can be done under which scenario.

Situation-Specific Dialogue Models: Next, for each case, the tool can automatically generate a dialogue model from the task model. After that, inter-model links are automatically inserted between the dialogue model states and the task model tasks that are allowed for each individual state. The nodes of the dialogue model (states) of the different dialogue models are connected to denote which states can alter the situation between them.

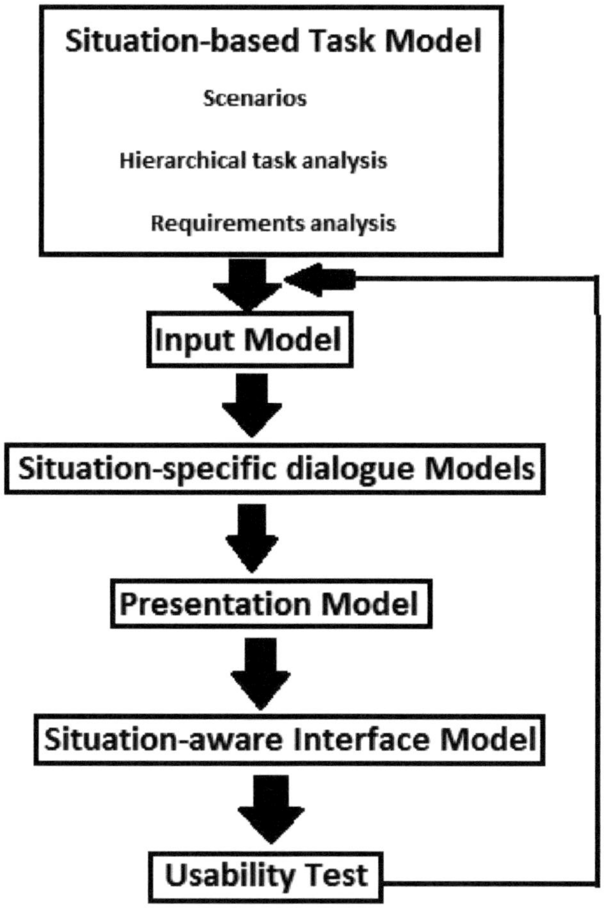

Figure 4.
Situation-aware Design Process for User Interface.

Presentation Model: Designers need to compose abstract UI components to provide the interface model with information about how the interaction should be interpreted to the user, and connect these to the relevant tasks for each node of the presentation model. In order to organize presentation components for layout purposes, the presentation model nodes may be arranged hierarchically. There are many abstract UI components for the designer to choose from, such as static, data, option, navigation control, hierarchy, and custom widget. Finally, it is possible to organize the UI components and arrange them into a hierarchical structure.

Situation-aware Interface Model: A situation-aware interface model benefits from the aggregate of all the models.

Usability evaluations: In order to assess and enhance the performance of the graphical interface of the models, usability tests are then conducted.

7. Current challenges and envisioning the future

Today, the development of intelligent user interfaces causes challenges equivalent to those encountered in Artificial Intelligence. It is well known that the optimal user interface is not the actual one in the area of human computer interaction. Currently, however, it is important to provide an intermediate between the wishes of the individual and the application of such intentions. It doesn't matter how elegant a user interface is, it will still be there to provide the user with a mental workload.

The potential machine, according to A. van Dam or T. Furness [42], would be a great butler who knows my setting, my preferences and my character, and without needing clear orders, gets ahead of my needs in a subtle way. When the user communicates with this butler, movements, facial expressions [43, 44] and other means of human speech, such as designing drafts, will mostly talk about the interaction. A goal of Artificial Intelligence from its very beginning, 50 years ago, is to be able to provide objects that facilitate learning, development or connectivity peer to peer with an individual.

To actually come true, by applying computer vision techniques or speech recognition to perceive and understand natural language, certain agents may be able to translate gestures and emotions. The inference engine for such agents would be Artificial Intelligence and knowledge-based technologies. They would do so in a friendly way when these agents speak with the user, and potentially modulate their voice according to the mood of the user at a given point in time.

The complexities of this technology can be separated into three areas: input, inference and output and, more precisely, the analysis of human language expressions, the depiction and control of knowledge of the world and, ultimately, the perception of human beings as social beings.

In [45], a virtual companion-based interface was proposed by the authors to simplify the mobile interface for the elderly. This interface shows menus through a virtual agent as per the scenario and the request of users and animates information in a 3-dimensional layout. To collect input from users, the authors used speech recognition technologies from smartphones and wearable devices. To imagine suitable actions based on user feedback, virtual avatars were chosen. The authors predict that this mobile interface might be the future of the next user interface for smartphones.

In the treatment and healing process, IUI is often used in tasks of determining the functional status of an individual. The performance of functional state estimation is dependent on combined data from the accelerometer and the EEG [46].

8. Conclusions

Computer products today have the potential to provide information to us, to entertain us or to make our lives easier, but whether the user interface provided is limited or difficult to use, they may also slow down our job. A vision of how various approaches from different fields, including Artificial Intelligence, Software Engineering and Computer-Human Interaction, have been added over the years to help establish a successful user interaction experience, increasing the overall usability of the device, has been seen in this chapter. In [47] the authors explore the community's evolving tacit viewpoint on intelligence characteristics in intelligent user interfaces and provide suggestions for expressing one's own perception of intelligence more clearly.

User interfaces can be as normal in the future as listening to a human. We should also note that Human-Computer Interaction's main aim is to reduce the mental distance between the user and his activities, dimming the interface until it becomes unseen. None the less, to get to the stage where we can provide users with virtual interfaces, there is always a long way to go. Computer-based automated interpretation of human sentiment and effect is generally predicted to play a significant role in future Intelligent User Interfaces, as it carries the potential of supplying immersive applications with emotional intelligence [48].

Another key of the study is that adding context to input in situation-aware systems results in automatic adjustment in awareness of the situation and action list, making the UI adapt to the actual operator's unique needs. In addition, adaptation can only actually happen where the context and climate adjustment is sufficiently important to result in the transition between two possible user interface statuses. Adapting the interface to the actual scenario as defined in this prototype and providing reusable tasks with a reduced number of commands, clicks and choices decreases the operator's cognitive burden and hence encourages interactions [49].

Author details

Nauman Jalil
NDSU, Fargo, USA

*Address all correspondence to: nauman.jalil@gmail.com

References

[1] ISO 9241-11. Ergonomic requirements for office work with visual display terminals (VDTs) - Part 11: Guidance on usability, 1998.

[2] NWIABU, N., ALLISON, I., HOLT, P., LOWIT, P. and OYENEYIN, B., 2012. User interface design for situation-aware decision support systems. In: IEEE Multi-disciplinary Conference on Cognitive Methods in Situation Awareness and Decision Support. 6-8 March 2012. Piscataway, New Jersey: IEEE. Pp. 332-339.

[3] Pavel Andreevich Samsonov. 2016. Improving Interactions with Spatial Context-aware Services. In Companion Publication of the 21st International Conference on Intelligent User Interfaces (IUI '16 Companion). Association for Computing Machinery, New York, NY, USA, 114-117. DOI:https://doi.org/10.1145/2876456.2876459

[4] Shumin Zhai. 2017. Modern Touchscreen Keyboards as Intelligent User Interfaces: A Research Review. In Proceedings of the 22nd International Conference on Intelligent User Interfaces (IUI '17). Association for Computing Machinery, New York, NY, USA, 1-2. DOI:https://doi.org/10.1145/3025171.3026367

[5] Philip Quinn and Shumin Zhai. 2016 A Cost-Benefit Study of Text Entry Suggestion Interaction. In Proceedings of the 2016 CHI Conference on Human Factors in Computing Systems (CHI '16).

[6] Shiri Azenkot and Shumin Zhai, Touch Behavior with Different Postures on Soft Smart Phone Keyboards, In Proceedings of MobileHCI (MobileHCI '12). ACM, New York, NY, USA, 251--260.

[7] Xiaojun Bi, Yang Li, and Shumin Zhai. 2013. FFitts law: modeling finger touch with Fitts' law. In Proceedings of the SIGCHI Conference on Human Factors in Computing Systems (CHI '13).

[8] Andrew Fowler, Kurt Partridge, Ciprian Chelba, Xiaojun Bi, Tom Ouyang, and Shumin Zhai. 2015. Effects of Language Modeling and its Personalization on Touchscreen Typing Performance. In Proceedings of the 33rd Annual ACM Conference on Human Factors in Computing Systems (CHI '15).

[9] Per-Ola Kristensson, Shumin Zhai, SHARK2: a large vocabulary shorthand writing system for pen-based computers, Proceedings of the 17th annual ACM symposium on User interface software and technology (UIST 2014)

[10] Philip Quinn & Shumin Zhai, Modeling Gesture-Typing Movements, Human-Computer Interaction, Pages 1--47, Published online: 26 Aug 2016

[11] Shumin Zhai and Per Ola Kristensson. 2012. The word-gesture keyboard: reimagining keyboard interaction. Commun. ACM 55, 9 (September 2012), 91--10

[12] Andrew Fowler, Kurt Partridge, Ciprian Chelba, Xiaojun Bi, Tom Ouyang, and Shumin Zhai. 2015. Effects of Language Modeling and its Personalization on Touchscreen Typing Performance. In Proceedings of the 33rd Annual ACM Conference on Human Factors in Computing Systems (CHI '15).

[13] Mitchell Gordon, Tom Ouyang, and Shumin Zhai. 2016. WatchWriter: Tap and Gesture Typing on a Smartwatch Miniature Keyboard with Statistical Decoding. In Proceedings of the 2016 CHI Conference on Human Factors in Computing Systems (CHI '16).

[14] Per-Ola Kristensson, Shumin Zhai, SHARK2: a large vocabulary shorthand writing system for pen-based computers, Proceedings of the 17th annual ACM symposium on User interface software and technology (UIST 2014)

[15] Zhai, S., Kristensson, P.O., Gong, P., Greiner, M., Peng, S., Liu, L. Dunnigan, A., ACM CHI 2009 Conference on Human Factors in Computing Systems Extended Abstracts. pp. 2667--2670.

[16] Shumin Zhai and Per Ola Kristensson. 2012. The word-gesture keyboard: reimagining keyboard interaction. Commun. ACM 55, 9 (September 2012), 91--10

[17] Shumin Zhai and Per-Ola Kristensson. 2003. Shorthand writing on stylus keyboard. In Proceedings of the SIGCHI Conference on Human Factors in Computing Systems (CHI '03).

[18] Jianlong Zhou, Jinjun Sun, Fang Chen, Yang Wang, Ronnie Taib, Ahmad Khawaji, and Zhidong Li. 2015. Measurable Decision Making with GSR and Pupillary Analysis for Intelligent User Interface. ACM Trans. Comput.-Hum. Interact. 21, 6, Article 33 (January 2015), 23 pages. DOI:https://doi.org/10.1145/2687924

[19] Paulo Salem. 2017. User Interface Optimization using Genetic Programming with an Application to Landing Pages. Proc. ACM Hum.-Comput. Interact. 1, EICS, Article 13 (June 2017), 17 pages. DOI:https://doi.org/10.1145/3099583

[20] D. Te'eni, J. Carey and P. Zhang, Human Computer Interaction: Developing Effective Organizational Information Systems, John Wiley & Sons, Hoboken (2007).

[21] B. Shneiderman and C. Plaisant, Designing the User Interface: Strategies for Effective Human-Computer Interaction (4th edition), Pearson/Addison-Wesley, Boston (2004).

[22] J. Nielsen, Usability Engineering, Morgan Kaufman, San Francisco (1994).

[23] D. Te'eni, "Designs that fit: an overview of fit conceptualization in HCI", in P. Zhang and D. Galletta (eds), Human-Computer Interaction and Management Information Systems: Foundations, M.E. Sharpe, Armonk (2006).

[24] Jaquero V.L., Montero F., Molina J., Gonz´lez P. (2009) Intelligent User Interfaces: Past, Present and Future. In: Redondo M., Bravo C., Ortega M. (eds) Engineering the User Interface. Springer, London. https://doi.org/10.1007/978-1-84800-136-7_18.

[25] Maybury, M. Intelligent user interfaces: an introduction. In Proceedings of the 4th international Conference on intelligent User interfaces (Los Angeles, California, United States, January 05 - 08, 1999). IUI '99. ACM Press, New York, NY, 3-4, 1999.

[26] López Jaquero, V., Montero, F., Molina, J.P., González, P. Fernández Caballero, A. A Seamless Development Process of Adaptive User Interfaces Explicitly Based on Usability Properties. Proc. of 9th IFIP Working Conference on Engineering for Human-Computer Interaction jointly with 11th Int. Workshop on Design, Specification, and Verification of Interactive Systems EHCI-DSVIS'2004 (Hamburg, July 11-13, 2004). LNCS, Vol. 3425, Springer-Verlag, Berlin, Germany, 2005.

[27] Paternò, F. Model-Based Design and Evaluation of Interactive Applications. Springer. 1999.

[28] Wahlster, W., Maybury, M. An Introduction to Intelligent User Interfaces. In: RUIU, San Francisco: Morgan Kaufmann, 1998, pp. 1- 13.

[29] G. Riva, F. Vatalaro, F. Davide and M. Alaniz, Ambient Intelligence: The Evolution of Technology, Communication and Cognition towards the Future of HCI, IOS Press, Fairfax (2005).

[30] Puerta, A.R. A Model-Based Interface Development Environment. IEEE Software, 1997.

[31] Eisenstein, J., Vanderdonckt, J., Puerta, A. Applying model-based techniques to the development of UIs for mobile computers. Intelligent User Interfaces 2001: 69-76.

[32] Maybury, M. Intelligent user interfaces: an introduction. In Proceedings of the 4th international Conference on intelligent User interfaces (Los Angeles, California, United States, January 05 - 08, 1999). IUI '99. ACM Press, New York, NY, 3-4, 1999.

[33] Puerta, A., Micheletti, M., Mak, A. The UI pilot: a model-based tool to guide early interface design Full text. Proceedings of the 10th international conference on Intelligent user interfaces. San Diego, California, USA. 215 –222, 2005.

[34] Lang, M., Fitzgerald, B. Hypermedia Systems Development Practices: A Survey IEEE Software. Volume 22 , Issue 2 (March 2005). 68 – 75. 2005. IEEE Computer Society Press Los Alamitos, CA, USA.

[35] Paternò, F. Model-Based Design and Evaluation of Interactive Applications. Springer. 1999.

[36] Vanderdonckt, J., Bodart, F. Encapsulating Knowledge for Intelligent Automatic Interaction Objects Selection. In ACM Proc. of the Conf. On Human Factors in Computing Systems INTERCHI'93 (Amsterdam, 24-29 April 1993), S. Ashlund, K. Mullet, A. Henderson, E. Hollnagel & T. White (Eds.), ACM Press, New York, 1993.

[37] López Jaquero, V., Montero, F., Molina, J.P., Fernández-Caballero, A., González, P. Model-Based Design of Adaptive User Interfaces through Connectors. Design, Specification and Verification of Interactive Systems 2003, DSV-IS 2003. In DSV-IS 2003: Issues in Designing New-generation Interactive Systems Proceedings of the Tenth Workshop on the Design, Specification and Verification of Interactive Systems. J.A. Jorge, N.J. Nunes, J. F. Cunha (Eds). Springer Verlag, LNCS 2844, 2003. Madeira, Portugal June 4-6, 2003.

[38] Fischer, G. User Modeling in Human-Computer Interaction. User Modeling and User-Adapted Interaction, Kluwer Academic Publishers, 2000.

[39] Horvitz, E., Breese, J., Heckerman, D., Hovel, D., Rommelse, K. The Lumiere Project: Bayesian User Modeling for Inferring the Goals and Needs of Software Users. Proceedings of the Fourteenth Conference on Uncertainty in Artificial Intelligence, Madison, WI, July 1998, Morgan Kaufmann: San Francisco, pp. 256-265.

[40] Bevan, N. Quality in Use: Meeting user needs for quality. Journal of Systems and Software 49(1): 89-96, 1999.

[41] Molina, J. P., García, A.S., López Jaquero, V. and González, P. Developing VR applications: the TRES-D methodology. First International Workshop on Methods and Tools for Designing VR applications (MeTo-VR), in conjunction with 11th International Conference on Virtual Systems and Multimedia (VSMM'05), Ghent, Belgium, 2005.

[42] Earnshaw, R., Guedj, R., van Dam, A. and Vince, J. Frontiers in Human-Centred Computing, Online Communities and Virtual Environment. Springer-Verlag, 2001.

[43] Picard, R. Affective Computing, MIT Press, 1997.

[44] Prendinger, H. and Ishizuka, M., 2005. The Empathic Companion: A Character-Based Interface that Addresses Users' Affective States. In Applied Artificial Intelligence, Vol. 19, pp. 267-285.

[45] S. Noh, J. Han, J. Jo and A. Choi, "Virtual Companion Based Mobile User Interface: An Intelligent and Simplified Mobile User Interface for the Elderly Users," 2017 International Symposium on Ubiquitous Virtual Reality (ISUVR), Nara, 2017, pp. 8-9, doi: 10.1109/ISUVR.2017.10.

[46] A. M. Syskov, V. I. Borisov and V. S. Kublanov, "Intelligent multimodal user interface for telemedicine application," 2017 25th Telecommunication Forum (TELFOR), Belgrade, 2017, pp. 1-4, doi: 10.1109/TELFOR.2017.8249439.

[47] Sarah Theres Völkel, Christina Schneegass, Malin Eiband, and Daniel Buschek. 2020. What is "intelligent" in intelligent user interfaces? a meta-analysis of 25 years of IUI. In Proceedings of the 25th International Conference on Intelligent User Interfaces (IUI '20). Association for Computing Machinery, New York, NY, USA, 477-487. DOI:https://doi.org/10.1145/3377325.3377500

[48] Björn W. Schuller. 2015. Modelling User Affect and Sentiment in Intelligent User Interfaces: A Tutorial Overview. In Proceedings of the 20th International Conference on Intelligent User Interfaces (IUI '15). Association for Computing Machinery, New York, NY, USA, 443-446. DOI:https://doi.org/10.1145/2678025.2716265

[49] Karray, Fakhri & Alemzadeh, Milad & Saleh, Jamil & Arab, Mo Nours. (2008). Human-Computer Interaction: Overview on State of the Art. International Journal on Smart Sensing and Intelligent Systems. 1. 137-159. 10.21307/ijssis-2017-283.

Chapter 9

Application of Artificial Intelligence in User Interfaces Design for Cyber Security Threat Modeling

Jide Ebenezer Taiwo Akinsola, Samuel Akinseinde, Olamide Kalesanwo, Moruf Adeagbo, Kayode Oladapo, Ayomikun Awoseyi and Funmilayo Kasali

Abstract

In recent years, Cyber Security threat modeling has been discovered to have the capacity of combatting and mitigating against online threats. In order to minimize the associated risk, these threats need to be modelled with appropriate Intelligent User Interface (IUI) design and consequently the development and evaluation of threat metrics. Artificial Intelligence (AI) has revolutionized every facet of our daily lives and building a responsive Cyber Security Threat Model requires an IUI. The current threat models lack IUI, hence they cannot deliver convenience and efficiency. However, as the User Interface (UI) functionalities and User Experience (UX) continue to increase and deliver more astonishing possibilities, the present threat models lack the predictability capacity thus Machine Learning paradigms must be incorporated. Meanwhile, this deficiency can only be handled through AI-enabled UI that utilizes baseline principles in the design of interfaces for effective Human-Machine Interaction (HMI) with lasting UX. IUI helps developers or designers enhance flexibility, usability, and the relevance of the interaction to improving communication between computer and human. Baseline principles must be applied for developing threat models that will ensure fascinating UI-UX. Application of AI in UI design for Cyber Security Threat Modeling brings about reduction in critical design time and ensures the development of better threat modeling applications and solutions.

Keywords: artificial intelligence, cyber security, human computer interaction, intelligent user interface, machine learning, threat model, user experience, user interface

1. Introduction

Cyber Security protect Information Technology (IT) assets, devices, data, programs, and networks from digital attacks or damage using processes, technologies, and practices to mitigate unauthorized access [1, 2]. Cyber security employs a variety of security concepts such as threat modeling to safeguard user's asset, environment and organization from being attacked. Threat modeling attempts to enhance cyber security by hypothetically analyzing, itemizing, and prioritizing the potential threats using the attacker's point of view [3]; with an accurate and deep understanding of threats to enable risk evaluation, and subsequently prioritize mitigation against an attack. Neglecting the security designers in cyber security and focusing only on human factors in cyber attacks [4] give room for wide penetration to sensitive systems through social engineering. Consequently, this could pose threats to physical machines and human manipulation via social engineering which is the most important aspect of information security. This ascertains the fact that the issue of cyber attack is progressively becoming a threat that militarizes information technology assets.

Cyber security cuts across technical and social matters and this need to be given priority as there have been a paradigm shift in the movement of businesses and government activities to the online platforms [2]. The complexities and interdependency on the technological system, poses a daunting threat to security engineers as devices and infrastructure are networked in socio-technical environment in which they operate. Therefore, the use of Application Program Interface (API) without the system engineer's control greatly undermines the system properties [5]. This gives rise to the creation of different laws with various provision of Cyber Security data protection, and privacy laws that operates from the inception with over 142 countries [6]. The provision of data protection laws and various security concepts such as threat modeling will be efficiently and effectively put into use if priority is also given to User Experience (UX) through the integration of intelligence in User Interface (UI) design. If UX is seen as having a huge effect on UI design, it could assist security experts in the development of intelligent threat modeling tools that can detect anomalies and make users understand the associated threats in every Human Computer Interaction (HCI).

1.1 Cyber security threat modeling

Threat Modeling (TM) is the basic building block for the establishment of secure systems. Threat modeling is a proactive move towards recognizing potential security concerns and an approach to risk assessment [7]. The success of any cyber security threat model is hinged on the development of an Intelligence System using Artificial Intelligence (AI) paradigms with keen attention on User Interface (UI) that takes into cognizance user interaction. According to [8], particular attention must be paid on what to offer to users as being regarded as a system/tool or an agent/assistant while considering Intelligent User Interface (IUI). IUI must be an Interactive System. Threat modeling must incorporate AI for efficient Human Computer Interaction (HCI) by leveraging on Machine Learning (ML) algorithms to develop a UI with automated capability and improved usability that guarantees high performance. Machine learning algorithm selection for model building can be subjective [9]. Hence, the need for a Multi-Criteria Decision Making (MCDM) approach for optimal algorithm selection [10] is desirable. The threat modeling tools must include components that perceive, learn, interpret, reason and decide; with emphasis on the decision about what is constituted as a threat or not.

Cyber threats lead to risks. Risks in cyber security are as a result of technical vulnerabilities and degradation of fundamental operational practices over time. These practices change organization from safe and sound practices region to a state where an attack will be successful with a greater organizational adverse effect [11]. The security teams investigate threat modeling to assign priority to threats to ensure that attention and resources are spread efficiently. The different levels of priority ensure correct threat mapping resulting from efficient mitigation. Threat modeling often assist security teams in maintaining safety against possible emerging threats as it is conducted frequently [7]. This frequency can be improved by applying intelligence in the User Interface (UI) design so that the cyber threats can be appropriately modeled using machine learning algorithms. Areas where there is a lack of protection in the adopted tools can also be clarified by modeling threats and facilitating team decision making where the component is appropriate. Krishnan [12] suggested that Intelligent User Interface (IUI) will also be helpful in the prioritization of current applications based on the predicted threat effects and magnitude.

2. Human machine interfaces

Human Machine Interfaces (HMI) deals with the study of two-way transmission of records among users and machines, which establishes a relationship with the user and create user experience [13]. It is natural for individuals to be proud of whatever they produce, no matter how simple it is or complicated as an HMI. Poor HMI designs have been seen as variables contributing to anomalous circumstances and thus exposing the design to cyber security threats. If the principles and practices of HMI are followed, the following are expected to be achieved for the design to be user-centered design:

1. Need to consider the user's aim, obligations with competence

2. Need to consider the mode of processing information by the users and making of decisions.

3. Need to keep the user on top of things and awareness of system condition.

If all these principles and practices are followed, they will improve the cyber security process, but regrettably, various HMIs designs were created with little human input, hence they are technology-centered and are not achieving the desired results. If the aim and the responsibilities of the users are considered with competence, attack threats will reduce. Also, consideration of the user's mode of processing information and decision making must be built into interface design. The successful application of baseline principles leads to trust and the level of trust the users have in a design will govern how they use it. For instance, if a user does not trust a site, he/she will not supply credit card details. If these principles are violated it can lead to credit card hacking. **Table 1** shows the relationship between poor HMI and effective HMI.

Poor HMI	Effective HMI
Piping and Instrumentation Drawing (PID) representation	Layout consistent with operators' model of the process (not a PID)
Presentation of raw data as numbers (temperatures, pressures, etc.)	Depiction of process status and values as information, not numbers
No trends	Key Performance Indicators (KPI) as trends
Bright colors, 3-D shadows	Gray backgrounds, low contrast
Color coding of piping and vessel contents	Consistent visual and color coding
Measurement units in large, bright text	Measurement units in low contrast lettering, if used at all

Table 1.
Characteristics of poor and effective HMI.

Intelligent HMI or HCI can help reduce security problem. A typical example of a security threat in the cyber space is phishing. Many phishing web pages are merely clones of actual sites with minor skewed or masqueraded elements in certain instances. These phishing sites' properties have made Intelligent HMI or HCI problematic for system users as well as several anti-phishing strategies to spot them. Attackers have been able to respond rapidly to anti-phishing initiatives that limits the efficacy of phishing attempts and defend unsuspecting users. Despite the impressive strides made by anti-phishing systems, in recent years, this assault remains one of the most successful. An IUI can be used to reduce the efficacy of phishing attacks and improve consumer understanding of associated threats. Using AI, the interface not only informs the user that there is a phishing attack but also discuss reasons why the website is a phishing site. Such a system is proposed in [14].

2.1 User Interface

User Interface (UI) allows the user to control software applications or hardware devices thus allowing the user have an interaction with the software or hardware of any device especially computing devices. User interfaces are available for both hardware and software devices. A typical example of a hardware device with user interface is the remote control; it has several buttons and sometimes screen to display some basic information. However, the buttons can be used by the user to tell the hardware what to do or what operations to perform. For instance, the use of a keyboard and mouse, each of which has its own user interface, to run a software program. Similarly, through the on-screen menus, a program graphical user interface, can be used to operate a digital camera. According to [15] the aim of a successful user interface, regardless of the program, is to be user-friendly.

UI focus on the looks and styles and it serves as the access point where the user interacts with the designs and system functionalities. UI comes in three forms such as Graphical User Interfaces (GUI), Voice-Controlled User Interfaces (VUI) and Gesture- Based User Interfaces (GbUI) [16]. The GUI allow users to interact with the visual representations of a digital control panel or system. Typical examples are the computer desktops and mobile phone screens. Unlike the GUI, the VUI allow users to interact with the systems and components via voices and speeches. Typical examples are the Google Assistant on Google devices, Siri on IoS devices and Alexa on Amazon devices. The VUI are rampant with more intelligent systems as there is need for a high level of voice recognition and speech processing. For Gesture- Based User Interfaces, the user gets more engaged in the three-dimensional (3-D) space and uses gestures, most especially hand gestures, in interacting with the system. These types of interfaces are mostly used in virtual reality spaces. Three key elements in UI include input control, navigational control and the informational component. Integration of cyber threat related feedback to UI for automated sentiment analysis of the system will aid qualifying potentially new type of attacks or threats.

2.2 User experience

Unlike the UI, that focuses on interfaces, the User Experience (UX) describes and centers on the user's experience of an interaction with the system. Interaction tends to be broader than interfaces as it facilitates dialog and communication between the user and the computing device. The Interfaces are used to experience interactions. Three major dimensions in UX are the users, products and interactions. UX helps to provide a simple way to analyze user product interactions and what influences them. The experience derived from the user product interaction at a given time and in particular conditions is regarded as UX [17]. It is the individual perception that results from the use or anticipated use of a product/service or system.

UX deals with analysis, experimentation, creation, content, and prototyping, whereas UI is a method of visually leading the consumer through all devices via the interface of a product utilizing interactive elements including cyber threat modeling. To solve a dilemma, UX focuses on all that concerns the user's path during and after the modeling process. UI, on the other hand, is a method that focuses specifically on how the interface of the product look and work. UX is a whole environment that cannot be confined to the screen. On the contrary, UI is typically visual and screen-related content [18]. A deep learning or ML that has the ability to continuously re-train itself based on user activities can be applied to increasing UX in making decision automatically. Example of such is the fake news/hate posts detection model used by some social media platforms.

3. Artificial intelligence in user Interface design

Design focus with Artificial Intelligence paradigms in User Interface is referred to as Intelligent User Interface (IUI) design. AI is revolutionizing industries and changing the status quo, and the design of UI is not left out in its disruption. AI is being used to design tools for providing details to UI designers to aid their designs. AI tools for User Interface (UI) with focus on User Experience (UX) like Uizard and Airbnb's Design AI can turn design sketches into product prototypes [19]. The interfaces must be intelligent enough for adequate detection of cyber threats. An AI-based User Interface threat model will assure reduction in cyber risks. A mainstream example of IUI is the GMail Smart Compose that offers predictive text to complete sentences thus aiding in writing emails faster [20]. An Intelligent Virtual Assistants (IVA) or AI assistant is another example of AI in UI design generally called IUI. Some notable IVA are Amazon Alexa, Microsoft Cortana and ChatBots. Generally, AI can be applied to UI designs on various interfacing channels between human and machines especially when modeling cyber security threats. Channels like search engines (Yahoo search, Google search, Bing, etc) and content recommendation systems (Netflix, Youtube, Spotify, etc) are also significant examples of IUI. Data privacy and considerate integrations, among others are challenges encountered with AI in UI designs, and they must be factored in during the design process for cyber security treat modeling [21]. IUI is aimed at incorporating intelligent automated capabilities and Artificial Intelligence (AI) into Human Computer Interaction. They are human machine interfaces that aims at improving effectiveness and efficiency of Human Computer Interaction (HCI) through the use of reasoning, user models, domains and media such as GUI, VUI and Gesture-Based User Interfaces [22]. IUI sits between AI and HCI. **Figure 1** depicts the relationship between AI and HCI with consideration to IUI.

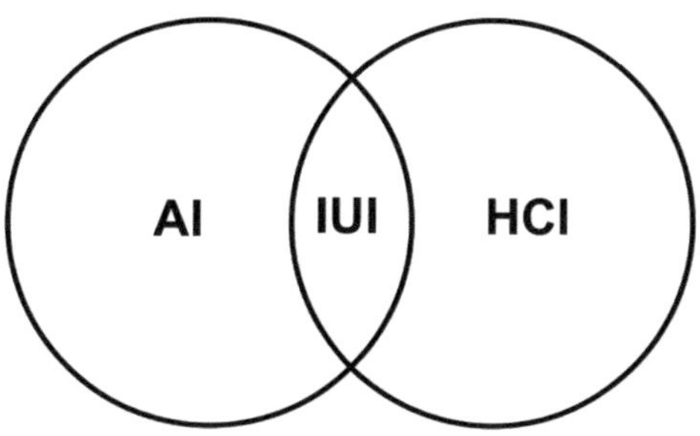

Figure 1.
Relationship between AI, HCI and IUI [23, 24].

HCI provides design techniques for user interfaces that are efficient and the AI components are used to embed intelligence into those interfaces. Basically, IUI is used to depict interfaces that generates some sort of output or exhibits some behavior in which the user interacting with the system considers the system intelligent [25]. A typical example is when a user clicks a wrong button and the interface is able to guide the user on the right button to click. Also, assistance to the user must be readily available through IUI.

Due to diverse models from differing fields that greatly influence user interfaces, the development of IUI is no different, as there are disciplines such as AI which contributes the intelligence in simulation approach to improve responsiveness; software engineering which allows for formal language definition, unified modeling approach, development life cycle; and the Human computer interaction that evaluates the user experience and avails techniques that can be used to create usable user interfaces [26]. In addition to these, there are other fields that still significantly contribute to the development of IUI which is depicted in **Figure 2**.

IUI tends to enhance the interaction between computer and human by bringing about novel approaches of communication and adapting the interface to the user using techniques of AI [27]. The concepts of IUI tend to the intersection of the area of AI and HCI in the field of IUI [28]. This is with a view to automate users' task while engaging in the threat assessment on cyber security [7]. Therefore, the application of intelligence in UI design will in no small measure bringing about better monitoring and controlling of activities that enhance better cyber security measures and threats modeling applications.

The advent of computers which primarily uses a mouse and keyboard as a medium of HCI has brought about a new dimension in the design of UI. This gives rise to the creative and innovative way of UI design for cyber security tools smartly through the use of AI. A new generation of UI which is called IUI keeps rising day by day and improving UX in HCI. IUI attempts the integration of intelligence in HCI that enables the automated capability of UI with the view to improve usability that enables high performance of HCI in cyber security software. It employs design and

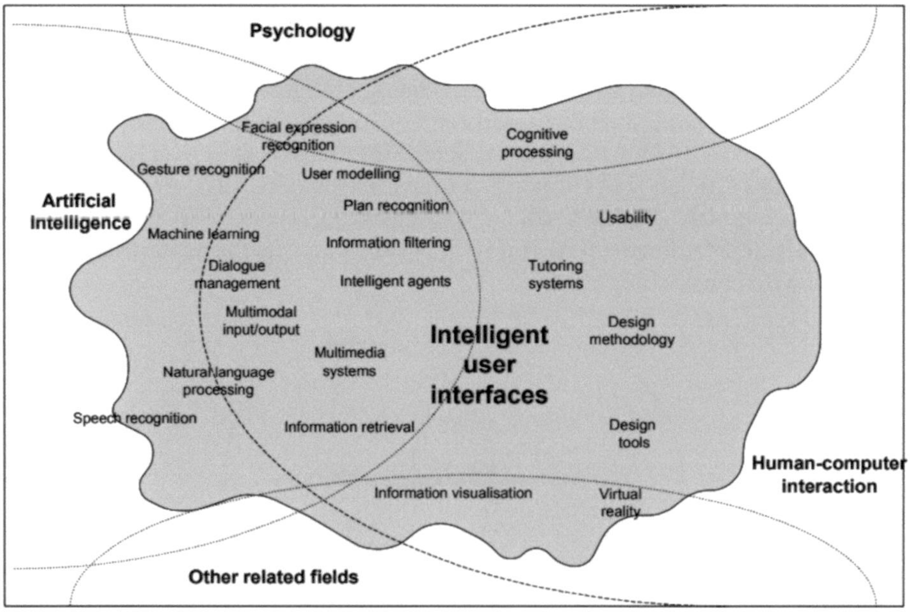

Figure 2.
Research fields of the intelligent user Interface [27].

implementations of AI components that perceive, learn, interpret, reason, and decide with a view to enhancing the capabilities of cyber security experts effectively [24] in decision-making. A different approach and tools can be employed in the implementation of IUI design. The choice of tools and approach is however depending on the goal and the area of application in the cyber security modeling tool. The strong effect of AI on UI/UX is that it helps fix challenges and avails insights and discoveries, however, humans have to recognize the challenges first. AI is therefore less likely to be pre-programmed and more likely to be built with technology so that it can fix the challenges by sustained learning and based on this learning, decisions can be more customized and user-centric. Data can be analyzed and reported in real time with AI.

Applications of machine learning (ML) that communicate directly with daily users are now progressively diverse and ubiquitous. ML is designed to allow a computer to learn about the past or the present and to forecast or predict the knowledge for unknown events in future [29]. The technology's sophistication and prevalence facilitates the belief that AI is the new UX; that is, AI would be the most effective way to boost the user experience [30]. This is because AI will permit UX designers to personalize contents through the application of machine learning algorithms to build intelligent models to mitigate cyber threats.

Typically, the development and design of UI occurs over the following phases as shown in **Figure 3**,

1. On a whiteboard or graphic tablet or even a sheet of tissue paper, designers of the program want to sketch their UI concepts by hand.

2. The designer uses a computer's wireframing method to build the same template again. This is a step that is repetitive.

3. The wireframes are converted into a functioning UI code by UI developers. Before the intended UI is designed, the developers and designers go through an iterative phase. This step is a procedure that is time consuming and tedious.

However, using AI, a handwritten design can be translated to a working UI as shown in **Figure 4**. This code generation usually go through a training phase and a sampling phase. The training phase is where the model is trained to identify images and state the relationships between them using algorithms like the Convolutional Neural Network (CNN) and Long Short-Term Memory (LSTM) and the sampling phase is more like the prediction phase where samples of UI to be translated will be supplied to the model so as to evaluate and fine tune it's performance.

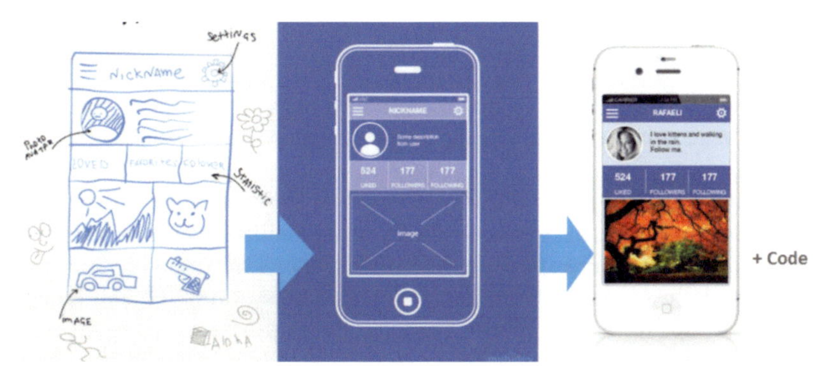

Figure 3.
UI development phases [31].

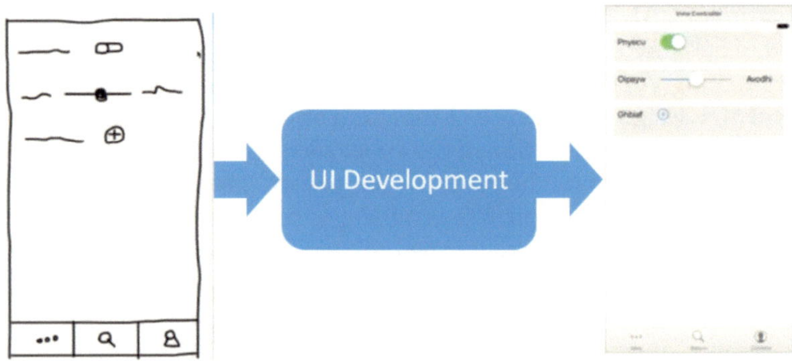

Figure 4.
Translating handwritten design into working UI [31].

The benefits of using AI in the design of UI include but are not limited to:

1. Rapid prototyping, improving iteration cycles, and ultimately better app development for both designers and developers. Critical project time will be saved on design projects when AI-based approach is adopted.

2. It would encourage designers and developers to concentrate on what matters most, adding value to end-users.

3. There will be very little or no barrier in application development as time to learn the use of UI design methods and time to code can be eliminated since everyone can draw UI on paper [31].

4. There is an elaborate interpretation and analysis of data for user-based customizations as vast amount of data can be collected. With AI, on the basis of data interpretation, you get the capacity to configure the interface according to consumer requirements.

5. Real-time learning and seamless adaptation can be achieved using AI in the design on UI. By combining deep learning that uses immense datasets to help draw a conclusion and this helps UI programmers easily build adaptable and improved interfaces.

6. It empowers the interface user. As AI continues to grow, consumers may have the ability to obtain more power over programs, eventually growing confidence and contributing to more usage.

7. It is possible to establish a deep bond with humans as AI systems gather and evaluate a large volume of data, resulting in better relationships.

AI systems can establish a deep bond with humans by creating relationships. Relationships foster where trust grows and this relationship could either be human to human or human to machine. One of the major reasons for the wide adoption of technology (especially AI) is the growing trusts in these systems. This reliance is strengthened by a repeated show of ability in achieving purpose, speed, accuracy, usability, security and privacy amidst other things [32]. By mining enormous data, through the use of deep learning, the trust between AI and humans can be strengthened by the continual generation of useful insights that fits into diverse purposes and ensure goals are achieved with speed, accuracy and security thereby increasing the usability of such systems. Considering the example of phishing site, the user's trust in the system is bound to increase because the system has protected the user from the phishing attack. Therefore, the user could rely on the AI system to detect phishing sites and help mitigate cyber threats. The more we interact with AI systems, the deeper the connection. For every interaction, data is generated and added to an individual's digital footprint on a system which AI feeds to improve its intellect with humans. AI technologies are also evolving rapidly and currently can analyze human emotions, sparking wild discussions into its unforeseeable future [33].

3.1 Empirical studies on intelligent user interfaces

The complexity and the number of cyber attacks is increasing [34] on a daily basis which poses a lot of threat to both public and private technological assets. Cyber security employ threat modeling techniques in mitigating attacks against technological asset by hackers. Threat modeling allows proactiveness and provides insight to risk evaluation and prioritize mitigation [35]. Formal methods have been found to be profound for performance evaluation compared to traditional methods [36]. As the cyber threat is becoming prominent, application of Artificial Intelligence (AI) in cyber security is another advancement in the technology employed by the experts in the field [34]. The manifestation of AI - enabled UI are Jarvis, Amazon Alexa, Neflix, IBM Watson, Nest Thermostat, Spotify and iRobot Roomba [37]. While advancement in technology to enhance security measure against cyber threat is expedient and important, User Experience (EU) also need to be given a high priority in the User Interface (UI) design to enable high usability and hassle-free workflow. As better security of system should not be tantamount to worse UX, the important factors for enhanced UX and UI in cyber security are balancing the security of system/software with UX, designing of UI based on human perception and minimizing the complexity of software integration into the existing network infrastructure [38]. A poorly designed UI reduces UX which leads to a user performing desired actions with difficulty [39]. An application of AI plays a key role in the IUI to mitigate problems that may arise from human interactions with machine. IUI is regarded as a subset of Human-Computer Interaction (HCI) research with a goal to use smart and current technology to improve HCI.

Applications of AI on UI can drastically improve the interaction between humans and computers. This makes it possible for computers to understand more human communication channels like body gestures, hand gestures, sounds, eye movements, lip-sync and other body motions [40]. This advanced communication with humans via AI has led to innovative solutions addressing human-computer communication barriers. An example is Conversational AI, which is an advanced platform for the widely used online service helpdesk. Conversational AI can analyze an individual's emotion and manage frustrations on the system by routing to different channels for better customer service [41]. With IUI, adaptive, personalized and responsive services can be provided to ensure the specific need of the user is met even when they are yet to realize it. With big data, data mining and deep learning algorithms, technologies that can drive these personalized services could be developed to work in synergy with intelligent environments to mitigate environmental challenges that may arise as a result of human computer interaction.

As the important features of IUI is to enhance the HCI, the following are current techniques used in IUI [27]:

1. **Getting of input from user intelligently:** This involve innovative way of getting user's input through several techniques such as recognition of face

and expression, processing of natural language, recognition and tracking of gesture, and tracking of gaze.

2. **Modeling of User:** This involves all communication techniques that allows the adaptability of human-machine interaction to different environment and users such machine learning, context awareness, among others.

3. **Generation of Explanation:** This covers the entire techniques that enables system explaining its result to users such as IU agents, speech output, feedback of tactile in a virtual environment

The application of intelligence in UI design is to improving users' experience that enables efficiency, effectiveness and user satisfaction using different approaches. This is achieved through the representation of reasoning or acting in accordance to a set of models such as user, dialog, domain, tasks, or speech. As shown in **Figure 5**, different models from different disciplines constitute the development of IUI. AI contributes the simulation of intelligent techniques to enhance the communication, software engineering- enabled notations, unified processes and formal languages while HCI deals on consideration pertaining to users. These combinations of models allow the creation of techniques that allow usable user interface.

IUI applies AI techniques to different input and output with a view to improving UX intelligently. This is achieved through reasoning, representation of knowledge, machine learning, adaptation, and adaptivity as exemplified in various applications such as email filter system, dialog system, email response system and so on [43]. As threat modeling provides answer to questions, "where, what, and how" [3], so also the metrics derive from these questions serve as input and output in AI to improve UX and integrate intelligence in UI designs.

There are two different methods of a user interface that can effectively be applied to intelligence user interfaces in cyber security threat modeling namely direct manipulation and indirect manipulation [7] as depicted in **Table 2**. Also, IUI have been anticipated as a way to overcome a number of the issues that direct manipulation interfaces cannot deal with and further highlighted three principles that can be applicable to threat modeling in cyber security, namely (i) control transparency and predictability (ii) Privacy and trust (iii) treating systems as fellow beings [47].

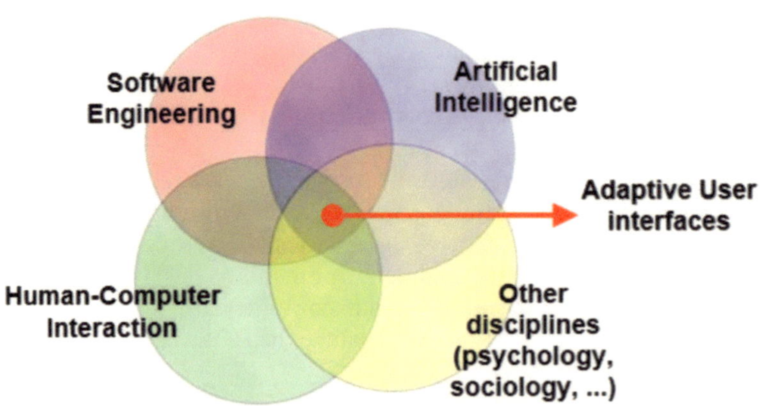

Figure 5.
Disciplines involved In the development of IUI [42].

IUI Design Analysis	Direct Manipulation	Indirect manipulation
Intelligibility: Supports the users to understand the interface action	High	Low
Predictability: Accuracy of algorithm	Low	High
Usability: For evaluating intelligence	High	High
Adaptation: Provision of intelligence in an interface	High	Low
[7, 27, 44–46].		

Table 2.
Comparative analysis of IUI.

Ehlert [27] also highlighted a number of issues that recent direct manipulations interfaces will not be able to handle for instance creation of tailored systems, filtering problems, provision of assistance on the use of complex and new programs, taking up the responsibilities from the users with other forms of interaction. The characteristics for developing intelligence in user interfaces designs that enhance the development of a threat model in cyber security are:

1. **Intelligent UI assist the user**: Assisting the user is most often seen as the key action that the intelligent entity performs.

2. **Intelligent UI adapt to the user and automate tasks**: Adaptation, automation, and interaction are the most well-known perspectives that specialists feature while portraying something as intelligent.

3. **Different UI concepts are intelligent in different ways**: Interfaces, systems, agents, and assistants are the most common entities to which researchers attribute intelligence [48].

3.2 Challenges in IUI

Lieberman [49] highlighted the following challenges about IUI for threat modeling in cyber security:

1. failure to give exceptional consideration to transparency and explanation.

2. evaluation of AI interfaces is challenging.

3. invest more in long-term interactions than the predictable interfaces.

4. self-mindfulness and appearance in projects have been a challenge

5. that predictability is not the solitary determinant of usability

Adaption of UI is a challenge for imparting intelligence which comprises the interpretation of user's events and correct prediction of the objectives. In addition, it is a common trade-off, deciding whether to use an easy approach so as to assist functionality. Finally, listed are the three major challenges:

1. Presentation (Human Computer Interaction stage of IUI),

2. Competence (focuses on Artificial Intelligence approaches or procedures)

3. Trust (on the IUI)

3.3 Comparative analysis of intelligent user interfaces

Intelligibility, predictability, usability and adaptation are four characteristics that must be consider while considering User Experience in the design of IUI as shown in **Table 2**.

The Intelligibility of IUI as it relates to how the users understand the interface actions was high when it comes to direct manipulation and low in indirect manipulation. This gives a better understanding of how and when intelligence can substantially improve the design practice (interaction) in the application of cyber security for threat model [47]. On the accuracy of algorithm which explain the predictability, there are many algorithms such as Dynamic Bayesian Network and Naïve Bayes can be used for user's prediction and techniques such as adaptation techniques that provide intelligence for enhancing the predictability of the UI and creating a lasting UX. The performance of these techniques decrease when the accuracy of algorithm by the system becomes low [44]. The usability principles for evaluating Intelligent User Interfaces (rather than direct manipulation systems) ensures that it does not mislead the user's expectation [48]. The provision of intelligence in a UI is highly reliable and cost efficient in direct manipulation than indirect manipulation as the authoring tool enable easy development and maintenance of the intelligent parts of the system [50] especially when it concerns threat modeling due to consideration for various risk metrics.

3.4 IUI design methods

A poorly designed interface can bring about inconvenience to its users. Just like every other product, IUIs need to be built carefully. The need for an IUI should be obvious from the analysis of the issue, instead of making an IUI simply because it is good. First of all, it should be determined whether a device needs an IUI or not because IUIs are typically more computer-intensive than conventional user interfaces. With a user interface, if the same performance can be achieved, why bother making a more complicated and expensive IUI? The final decision whether to build an adaptive process in interfaces or not lies in weighing the expense of deployment against the enhancement of user engagement [51]. If adaptive functionality is introduced in IUIs, the cognitive processing expected by the user will be reduced whereas, installation and management of an IUI would take time and computing resources. In designing IUIs, the following iterative refining steps below are used.

1. **Users, application, and environment review**: In every design process, the review and analysis stage is probably the most critical phase, but much more so in IUI design. There is need to analyze the typical system user, what roles the system is to play, and what method can execute in the design phase of a regular non-intelligent interface. Ideally, an IUI should be able to adjust to any consumer in any environment.

2. **Creation and installation of (prototype) interface methods:** The method of designing innovative strategies and metaphors for interaction is principally

one of imagination. Only getting out and seeking out fresh things and theories is the safest approach however, there are general interface architecture standards. IUI do not necessarily obey the general UI rules for example, user control, access power and consistency [52]. Other criteria, on the other hand, are best suited by IUI than by UI. The usage of natural language in IUI, for example, IUIs will speak the language of the user even more than conventional frameworks of UIs.

3. **Evaluation of the framework which was developed:** The criteria drawn up in the evaluation process should be complied and the feasibility of the prototype device should be studied while usability tests should be defined to assess this quality. These measures may include the number of errors, the completion time of the task, the user's interface attitude, etc. User satisfaction is a very significant but subjective usability criterion [52]. Since the consumer has to interact with the interface, there is need to know whether the design is nice and friendly to work with.

4. **Based on the evaluation results, corrections are made:** A range of concept changes would be made to the existing version depending on the issues found in the evaluation stage. A new round of design, execution, and assessment would then be initiated. Until the outcome is satisfactory, this iterative method will continue. The final interface methodology may be implemented into current user interface design tools if seen to be effective.

5. **Edit the tools for Designing interfaces:** There may also be a fifth stage in the design process, which is concerned with editing the tools for designing interfaces to add a modern approach or metaphor.

The following are some of the design approaches to the application of IUI for better user experience. **Table 3** shows the descriptions of various IUI design methods [53].

Design Methods	Description
Probability-driven State-charts	It uses state-charts to analysis users flow through different states. It creates a probability metrics of users expected actions with various features or events on an interface. This metrics then feed decisions on how the UI is personalized.
Decision Trees	This method utilizes contextual data to determine the choices of user groups. This analysis helps to make intelligent decisions on how different users interacts with a UI.
Prediction via shortest path algorithms	This approach uses weights to determine paths users will frequently take. Its process involves tracking the previous or source state, the next or target state, current event and time of the event to determine user's shortest paths.
Higher-order Markov model	This approach analysis multiple user transitions to predict user's decision on the next path to take. This analysis is focused on users previous or source paths and goes beyond the user's last path to its extended source paths to a destination.
Deep Reinforcement Learning (DRL)	This design approach does an extended prediction into expected user activities. DRL analysis itself towards the predictions of user's behaviors and it either regards itself if the predictions are correct or get punished if otherwise. This learning is done over-time to improve itself towards becoming a perfect user prototype with knowledge of expected actions.

Table 3.
IUI design methods.

3.5 AI versus UI design

The distinguishing factor between the traditional UI design and IUI design is the application of AI for better User Experience (UX). AI introduces a raw ability towards creating individualized interfaces through insights and discovery from understanding users' actions [54]. The user's expectations or intended activity on an interface is what AI feeds on to present a better user experience. The traditional UI is still relative to the design, but AI factors the user's interest to give them lasting User Experience (UX). With the traditional search engines, algorithms are rigid and yield similar search results for the same search input from different users, but AI evolves the search engines to generate smarter results tailored to a specific user. Keyboards is another rigid interface being gradually replaced by sensor-based interactions like body gestures, touchless hand gestures and sounds. Affective computing is another AI intervention, it factors human emotions, moods and expressions [55]. It's an emerging technology that enables a UI to understand and respond to human emotions in hopes of changing the rigid form of human interactions with computers.

Pertinent question must be asked when it comes to AI and UI design. What is the importance of AI to the design of User Interface? This question can only be answered with recourse to HCI. Every process in industrial design field of HCI and AI are intertwined from earliest stage in computer development with core relationship as intelligence even as at today [49]. The intelligence ushers in knowledge representation and interaction management with the application of machine learning algorithms in the design of IUI. The emergence of HCI has brought about a lot of innovations which led to technological development propelled by ubiquitous computing [56]. The idea behind HCI is making computers very easy to use and more helpful while AI is modeling human idea and exemplify those ideas into computer. These relationships bring about the creation of IUI with a high degree of usability. Therefore, AI unravels many difficulties being confronted with in User Interface design and answers many questions relating to User Experience. **Table 4** depicts HCI goals in relation to User Interface design and AI.

HCI Goals	User Interface Design	Artificial Intelligence
Collaboration	It gives attention on collaboration among remotely-located users	It unites the imagined joint effort between the user and the computer, as the computer takes a more lively function
User Input	It is more flexible as users can uninhibitedly blend significant level communication and give definite and explicit directions.	It can give a more flexibility to computer systems input.
Adaptation	It studies users and transfer the results at design time to designers and implementors.	It offers the capacity to move part of the loop comments from design time to run time.
Desired Work Practices	It describes the real-world work practices and assure the systems is constant with these practices.	It represents the work practices as workflow descriptions and document content (both procedural and declarative)
User Experience	It helps users to achieve their personal and work tasks and goals.	It gives opportunity to concretely implement the user's ideas directly in working systems.

Table 4.
AI versus UI design.

The application of AI in UI design is mainly a digital practice and will visually guide the user over the product's interface. There is a brighter future for integration of AI into UI, even though there are several AI algorithms with heuristic nature but these solutions (predefined model, making more established communication as less significant than current ones and excessive accuracy) can assist in the threat modeling for cyber security [44].

3.6 IUI design model

IUI design model depicts the representation of the UI and how the UX will be. The selection of design model is very important because this determines what features and services will be available in the user interface. Using a blueprint may be a major boost in the design process [57]. Some models have an overall architectural model that focuses on user modeling, whilst others propose a multi-modal input architectural model. Input is captured and later pre-processed from the keyboard, mouse, microphone, video, or probably some other input unit. Production entails case marking and other input elements that are interesting. The different modalities are fused and measured after each input modality has been evaluated.

Once all the knowledge required is accessible and revised, the framework must settle on the best choice for intervention. This is called adaptation in **Figure 6**, as some types of interface adaptation are typically selected. Assessment and adaptation sometimes take place concurrently using one inference engine for both, rendering the difference between the process of assessment and adaptation not very obvious. It is also important to produce the selected operation, which is achieved in the segment of performance production. Most IUI can be produced with this model or equipped with it. A general IUI model is shown in **Figure 6**.

3.7 AI-based UI and UX prototyping tools

AI is driving simplicity in UI and UX design, and the following are some of the top use cases.

1. **Uizard:** This tool enables for rapid prototyping of sketches. It converts hand-drawn wireframes into application prototype with working code.

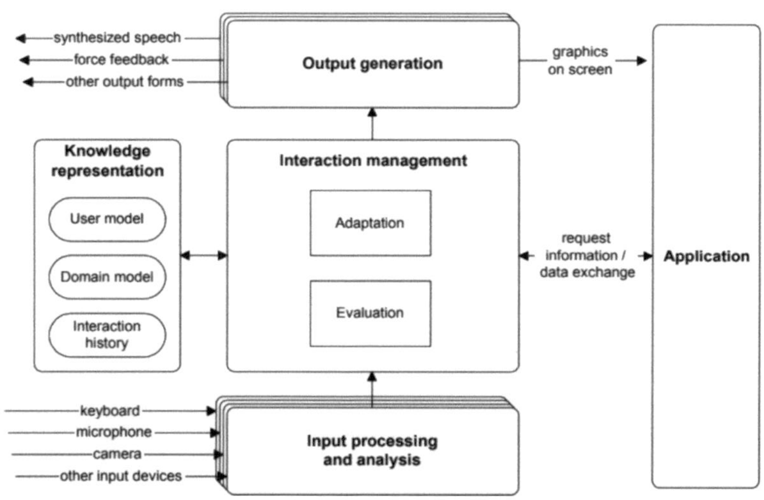

Figure 6.
Intelligent user Interface model [27].

Features	Uizard	Airbnb's Design AI	Mockplus	InVision	Balsamiq
Learning Curve	Minimal	Minimal	Minimal	Minimal	Minimal
Simplicity	High	High	High	Medium	High
Offline capabilities	No	No	No	Yes	Yes
Free to use (No trial)	Yes	No	Yes	Yes	No
Collaboration	No	No	Yes	Yes	Yes
Community Support	Low	Low	Medium	High	Medium
Supported Platforms	Web	Web	Mac / Win	Web	Mac / Win /Web
Code Generation	Yes	Yes	Yes	No	No
In Production	No	No	Yes	Yes	Yes
Developer Specs	Yes	No	No	Yes	No
Screen Transitions	Yes	Yes	Yes	Yes	No
Designs from Scratch	Yes	Yes	Yes	No	Yes
Active Usage	Low	Low	Low	Medium	Low

Table 5.
Comparative analysis of commonly used IUI design tools.

2. **Airbnb's Design AI:** This tool is currently in development and can easily convert design sketches to product prototype with working code.

3. **InVision**: This prototyping tool converts uploaded designs into interactive prototypes.

4. **Mockplus**: This is used for quickly creating interactive prototypes of applications.

5. **Balsamiq Mockups**: This tool easily designs wireframes of applications. The design process and generated wireframes are in sketch format. [58–61]. **Table 5** shows the comparative analysis of commonly used IUI Design Tools.

4. Cyber security threat models

The method of threat modeling is a collection of techniques used to construct a system abstraction, profile possible attacks, goals and procedures, including potential threats index that may occur [62]. A typical threat modeling process includes five components which are threat intelligence, asset identification, mitigation capability, risk assessment and threat mapping. Threat models require the application of Artificial Intelligence for prompt risk reporting. In the second quarter of 2018, the cost of cyber crime damage is projected to hit $6 trillions annually by 2021 and the vulnerabilities of the malware that target machine have been up to 151%. Therefore, Intelligence User Interface (IUI) is desirous for threat modeling. There can be external or internal threats, with catastrophic consequences. Attacks may completely bypass programs or leak confidential information that reduces the customer interest in the system's provider.

There are over a dozen mainstream threat models applied to Cyber Security. Threat models have been in existence since the late 1990s to help fortify cyber security. Meanwhile, [63] noted that irrespective of the availability of various threat models, they all follow five steps. First is a granular breakdown of infrastructure or application, then determining the threats, followed by its preventive measures, next is severity reduction (mitigations) and lastly, ranking of threats [35]. The predictability capacity of the machine learning algorithms is a major factor in threat modeling. There are twelve threat models, but the pioneering model is Microsoft's STRIDE that sprung-off in 1999 and gave rise to subsequent models. STRIDE results in a few false positives and is ideal for teams with little security expertise [64]. Other top models include DREAD, PASTA, LINDDUN, Trike, OCTAVE, CVSS, hTMM, Attack Trees, Persona Non-Grata (PnG), Security Cards, Quantitative TMM and VAST Modeling [65]. An analysis of threat models discovered that hybrid threat models handle more potential attacks than single models [62]. This analysis also recommended the PASTA model because it has a well-structured layout, and its implementation is attributed to sub-systems rather than the whole system. Another threat model worthy of note is the Persona Non-Grata (PnG), tailored to UX design, where users' behaviors and interactions with UIs are analyzed. It outputs a few false positives and it is ideal when the system's weakness is known else it becomes difficult to pinpoint likely threats. The Security Cards model is another endorsement ideal for unusual or advanced attacks and is dependent on brainstorming about possible intrusions. The major drawback of Security Cards is its high false positives [64]. **Table 6** shows the comparative analysis of the features of the most common threat modeling methods that are widely in use.

4.1 Cyber security threat modeling process

Threat modeling involve a number of processes and aspects for efficient mitigation. Failure to include these set of components might lead to incompletion in modeling thereby preventing proper threats prevention. The list of those components is as follows:

4.1.1 Threat analysis

This is called threat intelligence. This has to do with the granular breakdown of infrastructure or application. The section contains information on threat types, affected devices, monitoring mechanism, vulnerability exploitation tools and processes and attackers' motivations. Security analysts also gather and use online sources, proprietary solutions or security communications channels to access information on threat intelligence. This is used to improve the awareness and knowledge on emerging risks to determine the right course of action. Most importantly, it tries to understand the data flow across the system.

4.1.2 Identification of asset

This is called threat determination. Security teams need an in-house inventory of the components and data used, the location of those assets and the security procedures used. This inventory allows security teams to monitor identified vulnerabilities for their assets. It helps to gain insight for asset modifications from an inventory in real-time. For instance, warnings to the possibilities of attacks if assets are introduced, with or without allowed approval. This involves identifying all potential and current threats to the applications/systems.

Features Versus Models	Stride	Pasta	Security Card	Attack Tree	CVSS	QTMM
Documentation	Extremely High Documentation	High Documentation	Less Documentation	Less Documentation	Less Documentation	Extremely High Documentation
Technical Threat Identification	Highly Suitable	Highly Suitable	Highly Suitable	Highly Suitable	Highly Suitable	Highly Suitable
Non-Technical Threat Identification	Highly Suitable	Suitable	Suitable	Highly Suitable	Highly Suitable	Highly Suitable
General Threat Identification	Very Efficient	Very Efficient	Efficient	Moderately efficient	Very Efficient	Very Efficient
Time Consumption	High Time consuming	Extremely time consuming	Moderately time consuming	Moderately time consuming	Moderately time consuming	Moderately time consuming
Usage	Very Easy to Use	Difficult to use	Moderately easy to use	Easy to use	Easy to use	Easy to Use
Model Maturity	High Maturity	Medium Maturity	Low Maturity	Medium Maturity	Medium Maturity	High Maturity
Area of focus	Developer	Attacker / Application	Enterprise	Acceptable Risk	Attacker / Application	All Encompassing
Training / Usage Requirements	Required Less Training	Required More Training	Require Moderate Training	Require Moderate Training	Require Moderate Training	Require Moderate Training
Business Impact	Low	Extremely High	Low	Low	Medium	Medium
Threat Output	Threats properties, system entities, incidents and system limits	Threat management, enumeration, and scoring.	Target out-of-the ordinary threat (sophisticated threats)	System direction and Attack Targets	Scoring System and Severity Ranking	Threats properties, system entities, Scoring System and Severity Ranking
Security Properties	Extremely High	Very High	High	Very High	High	Extremely High
Areas of application	Software Industry and Engineering	Software Industry, Engineering and Banking	Production	Construction and Production	Software Industry and Banking	Software Industry and Project Management

Features Versus Models	Stride	Pasta	Security Card	Attack Tree	CVSS	QTMM
Threat Classification	Very Efficient	Efficient	Moderately Efficient	Efficient	Highly Efficient	Highly Efficient
Stakeholders Input / collaboration	Very High Collaboration	Extremely High Collaboration	Extremely High Collaboration	No Collaboration	No Collaboration	Very High Collaboration
Threat Prioritization	Medium Prioritization and Moderately Efficient	Extremely High Prioritization and Highly Efficient	High Prioritization and More Efficient	Medium Prioritization and Moderately Efficient	Extremely High Prioritization and Highly Efficient	Extremely High Prioritization and Highly Efficient
Reliability	Highly Reliable	Highly Reliable	Extremely Low Reliability	Moderately Reliable	Highly Reliable	Highly Reliable

Table 6.
Comparative analysis of threat modeling methods.

4.1.3 Mitigation capacity

This is called countermeasures. The capability of mitigation (that is, capacity of mitigation) usually apply to technologies for securing, identifying and responding to a particular form of threat but also could mean the security skills, know-how and processes of an enterprise. Assessing the current expertise will help decide whether additional resources are required to minimize a threat. For instance, there could be an initial degree of security against typical malware attacks if there is company-grade Anti-Viruses (AV). To compare the current AV signals with other detection capacities, for example, the security expert can decide if there is a need to invest more. This is centered on preventive measures. This involves analysis of current application cyber attacks, managing the damages done, and fortifying system security.

4.1.4 Evaluation of vulnerability

This is called mitigation or risk assessment. This step addresses identified threats with a focus on high-risk threats. Risk evaluations are related to asset inventories through threat intelligence. These resources are required to help security teams think about their systems' current state and develop vulnerability management strategies. Active device and solution monitoring can also provide risk evaluations. Penetration testing are, for example, effective in checking protection measures. This is to reduce severity.

4.1.5 Mapping of threats

Threat mapping is a method that traces the system's likely path to threats. It seeks to model how assailants can switch from resource to resource and help security teams predict where protections can be implemented or enforced more efficiently. It deals with ranking of threats according to their severity and potential damages to the application.

4.2 Predictive analytics of threat models

Machine Learning (ML) assist in the mathematical models construction which has the ability to explain and showcase complicated behavior without the need for programming [50]. These techniques have a way of improving the HMI in the interface design and further improve the intelligence of the design. The usage of ML in Human Machine Interface/Interaction (HMI) design is not very trivial. Moustakis and Herrmann [66] affirmed that the misperceptions about ML, inadequate familiarity with ML's latent capacity and research shortage are the main sources for declined utilization of ML in HMI design. Though, as of today ML has gained recognitions and advance applications can enhance user capability in UI design which helps in mitigating against cyber security threats. Integration of models that use logs to reactively discriminate transactions based on user' history is essential for cyber security threat modeling.

The challenge usually face during threat analysis is now being fazed away by applying AI coupled with machine learning algorithms, which feeds on data to detect abnormalities in systems. Predictive analysis is AI-driven by data and uses large data to understand malicious activities, identify patterns, and provide insights into potential attacks much quicker [67]. Standardization does not exist, and thus the choice of threat models is deterministic of the project needs like targeted risk area, allocated time, expertise, and stakeholder's involvement [68]. Furthermore, it is advantageous to apply threat models at the requirement and design stage of the project life-cycle [65] and the use of a well-formulated model for Software Development Life Cycle (SDLC) [69] for efficient threat modeling is desirous. Though AI sheds a positive light on cyber security, it also presents alarming intrusion possibilities for cyber criminals. This ordeal does not limit the impact of AI but rather re-enforces its significance, especially in cyber security threat modeling.

4.3 Baseline principles for intelligent user interfaces design

There are four baseline principles that are essential for designing an AI-enabled User Interface [70].

1. **Management of discovery and expectation**

   User expectation is very critical in the design of Intelligent User Interface so that false expectation can be avoided. Thus, the users will know the capability of the tools and its limitation as well as benefits expected with minimal input.

2. **Design to Forgive Mistakes**

   The AI is going to make errors. The user interface should be design so that users want to forgive any mistake they encounter from the tools. There should be capacity to understand response from natural language.

3. **Data Transparency and Customization**

   There should be transparency in data collection and allow users to customize with vivid dashboard for efficient monitoring. There should be provision for input from the users to alter what AI has learned apart from what has been programmed from the business logic.

4. **Privacy, Security, and User Control**

   Security must not be compromised in the design of IUI. Users confidence must be gained by ensuring privacy, security, and the ability for AI control. The user should trust the IUI with their personal data.

4.4 Application of AI in UI design for cyber security threat modeling

Threat Modeling in cyber security assists in examining current and potential vulnerabilities within a system, and has been an instrumental process against security threats [71]. Though favorable, this process has hardly warded off looming security threats that have kept a constant necessity for cyber security improvement, leading to conventional approaches on Threat Modeling. The introduction of AI into UI results in behavioral analysis, which can be applied to the second step of threat modeling called threat determination for forestalling cyber attacks. An effective AI approach tailored to vulnerability management has been on behavioral analysis on an attacker [72].

Threat Modeling incorporates AI to analyze various user interactions with interfaces and detect anomalies for potential attacks. Some software solutions or applications designed to detect cyber threats are Darkrace Immune System, a cyber security platform that uses AI to learn human interaction patterns on system interface for anomaly detection, Vectors' Cognito and Paladion [73]. Codesealer is another software application that provides UI security [74]. Another application is

Automated Virtual Agent for Truth Assessment in Real-Time (AVATAR), a United State government security screening tool designed to detect false information during user interaction with the system and is used for automated interviews at Airport checkpoints [75]. Some of the other applications of AI in cyber security are in the areas of spam filtering and malicious traffic detection. These are cyber threats which requires intelligent models to mitigate the attacks.

It is vital to note that AI has become a valuable tool for cyber criminals, thus reinforcing the significant practice of AI in UI designs [73]. AI application in UI is not left without drawbacks which are generally centred on the lack of pattern-driven dataset, computing and data resources. In addition, the introduction of AI attracts AI-modeled attacks like model evasion, data poisoning and data-stealing, though they can be managed by AI domain expertise with good security practices and safeguards.

5. Conclusion

Cyber security should employ a variety of security concepts such as threat modeling techniques to safeguard asset, environment and organization from being attacked. Five essential components of threat modeling process which are threat intelligence, asset identification, mitigation capability, risk assessment, and threat mapping are pivotal to mitigating security threat that are based on IUI design while taking into cognizance the baseline design principles. There must be proper application of different threat model methods at the requirement and design stages of the project life-cycle using a well-formulated Software Development Life Cycle (SDLC) model. The design of Intelligent User Interfaces requires incorporation of Artificial Intelligence (AI) to analyze various user interactions to detect anomalies such as phishing, spamming, fake news/hate posts, among others. To achieve this, predicative analytics must be institutionalized to further improve the intelligence of the design. Deep leaning can be used for sentimental analysis to determine expressions that can aid in qualifying potentially new type of attacks or threats. AI-driven User Interface (UI) design must be user-centered instead of technology-centered to give lasting UX which cannot be derived or experienced with the traditional UI Design. HCI provides design techniques for User Interfaces that are efficient and the AI components are used to embed intelligence into those interfaces. Therefore, appropriate IUI Design Methods, models coupled with AI-Based UI and UX Prototyping Tools are crucial for developing efficient cyber security threat modeling applications. This is highly essential because in the IUI design, direct manipulation method of design has low predictability which does not guarantee effective cyber threat mitigation.

Author details

Jide Ebenezer Taiwo Akinsola[1*], Samuel Akinseinde[2], Olamide Kalesanwo[3], Moruf Adeagbo[1], Kayode Oladapo[3], Ayomikun Awoseyi[1] and Funmilayo Kasali[4]

1 First Technical University, Ibadan, Nigeria

2 The Amateur Polymath, Lagos, Nigeria

3 Babcock University, Ilisan-Remo, Nigeria

4 Mountain Top University, Ibafo, Nigeria

*Address all correspondence to: akinsolajet@gmail.com

References

[1] DigitalGuardian, "What is Cyber Security ? Definition, Best Practices & More," *DATA PROTECTION 101*, 2020.

[2] M. Veale, I. Brown, and F. Getulio, "Cybersecurity," *J. Internet Regul.*, vol. 9, no. 4, pp. 1-22, 2020, doi: https://doi.org/10.14763/2020.4.1533 Published:

[3] A. Hayes, "A Guide to Easy and Effective Threat Modeling," *Application Security*, 2020. .

[4] Adams A. and Sasse M. A., "U SERS A RE N OT is therefore," *Commun. ACM*, vol. 42, no. 12, pp. 40-46, 1999, doi: doi.org/10.1145/322796.322806.

[5] B. Kostova, S. G.¨urses, and C. Troncoso, "Privacy Engineering Meets Software Engineering.," *On the Challenges of Engineering Privacy By Design*, vol. arXiv:2007. pp. 1-17, 2020, doi: http://arxiv.org/abs/2007.08613.

[6] G. Greenleaf and B. Cottier, "2020 ends a decade of 62 new data privacy laws," 2020.

[7] V. L. Jaquero, F. Montero, J. P. Molina, and P. González, "Intelligent User Interfaces: Past, Present and Future," *Eng. User Interface From Res. to Pract.*, no. September, pp. 1-282, 2009, doi: DOI: 10.1007/978-1-84800-136-7_18.

[8] D. Buschek, "What makes user interfaces intelligent? | by Daniel Buschek | UX Collective," 2020. https://uxdesign.cc/what-makes-user-interfaces-intelligent-9f63b27ca39 (accessed Dec. 26, 2020).

[9] J. E. T. Akinsola, O. Awodele, S. A. Idowu, and S. O. Kuyoro, "SQL Injection Attacks Predictive Analytics Using Supervised Machine Learning Techniques," *Int. J. Comput. Appl. Technol. Res.*, vol. 9, no. 4, pp. 139-149, 2020, doi: 10.7753/ijcatr0904.1004.

[10] J. E. T. Akinsola, O. Awodele, S. O. Kuyoro, and F. A. Kasali, "Performance Evaluation of Supervised Machine Learning Algorithms Using Multi-Criteria Decision Making Techniques," in *International Conference on Information Technology in Education and Development (ITED)*, 2019, pp. 17-34, [Online]. Available: https://ir.tech-u.edu.ng/416/1/Performance Evaluation of Supervised Machine Learning Algorithms Using Multi-Criteria Decision Making %28MCDM%29 Techniques ITED.pdf.

[11] T. R. McEvoy and S. J. Kowalski, "Deriving Cyber Security Risks from Human and Organizational Factors – A Socio-technical Approach," *Complex Syst. Informatics Model. Q.*, vol. 03, no. 18, pp. 47-64, 2019, doi: 10.7250/csimq.2019-18.03.

[12] S. Krishnan, "A Hybrid Approach to Threat Modelling," 2017. doi: 10.13140/RG.2.2.33303.88486.

[13] W. Dong *et al.*, "Soft human–machine interfaces: design, sensing and stimulation," *Int. J. Intell. Robot. Appl.*, vol. 2, no. 3, pp. 313-338, 2018, doi: 10.1007/s41315-018-0060-z.

[14] J. Aneke, C. Ardito, and G. Desolda, "Designing an Intelligent User Interface for Preventing Phishing Attacks," in *Lecture Notes in Computer Science (including subseries Lecture Notes in Artificial Intelligence and Lecture Notes in Bioinformatics)*, 2020, vol. 11930 LNCS, pp. 97-106, doi: 10.1007/978-3-030-46540-7_10.

[15] P. Christensson, "User Interface Definition," 2020. .

[16] IDF, "User Interface Design," *Interaction Design Foundation (IDF)*, 2020. https://www.interaction-design.org/literature/topics/ui-design (accessed Dec. 20, 2020).

[17] W. Chen *et al.*, "Development and Application of Big Data Platform for Garlic Industry Chain," *Comput. Mater. Contin.*, vol. 58, no. 1, pp. 229-248, 2019, doi: 10.32604/cmc.2019.03743.

[18] Guru99, "UX vs UI: 10 Most Important Differences You Must Know!," 2020. https://www.guru99.com/ui-vs-ux.html.

[19] H. Sarmah, "Top 5 AI-Based Prototyping Tools For UI And UX," 2019. https://analyticsindiamag.com/top-5-ai-based-prototyping-tools-for-ui-and-ux/.

[20] Google, "Use Smart Compose - Computer - Gmail Help," 2020. https://support.google.com/mail/answer/9116836?co=GENIE.Platform=Desktop&hl=en.

[21] S. T. Völkel, C. Schneegass, M. Eiband, and D. Buschek, "What is ' Intelligent ' in Intelligent User Interfaces ? A Meta-Analysis of 25 Years of IUI," in *in Proceedings of the 2020 Conference on Intelligent User Interfaces (IUI'20).*, 2020, pp. 1-20, [Online]. Available: https://arxiv.org/pdf/2003.03158.pdf.

[22] D. Bachmann, F. Weichert, and G. Rinkenauer, "Review of three-dimensional human-computer interaction with focus on the leap motion controller," *Sensors (Switzerland)*, vol. 18, no. 7. MDPI AG, Jul. 2018, doi: 10.3390/s18072194.

[23] V. Potluri, T. Grindeland, J. E. Froehlich, and J. Mankoff, "AI-Assisted UI Design for Blind and Low-Vision Creators," 2019.

[24] D. Sonntag, "Intelligent User Interfaces," in *ISMAR 2015 Tutorial on Intelligent User Interfaces*, 2015, no. May, pp. 1-24, doi: 10.1016/b978-0-08-028572-6.50016-x.

[25] M. Maybury, "Intelligent User Interfaces: An Introduction," 2019.

[26] P. Cybulski and T. Horbinski, "User Experience in Using Graphical User Interfaces of Web Maps," *ISPRS Int. J. Geo-Information*, vol. 9, no. 7, Jul. 2020, doi: 10.3390/ijgi9070412.

[27] P. Ehlert, "Intelligent User Interfaces: Introduction and Survey," 2003.

[28] M. Eiband, S. T. Völkel, D. Buschek, S. Cook, and H. Hussmann, "When People and Algorithms Meet : User-reported Problems in Intelligent Everyday Applications," in *24th International Conference on Intelligent User Interfaces*, 2019, pp. 96-106, doi: 10.1145/3301275. 3302262.

[29] J. E. T. Akinsola, M. A. Adeagbo, and A. A. Awoseyi, "Breast cancer predictive analytics using supervised machine learning techniques," *Int. J. Adv. Trends Comput. Sci. Eng.*, vol. 8, no. 6, pp. 3095-3104, 2019, doi: 10.30534/ijatcse/2019/70862019.

[30] K. Carmona, E. Finley, and M. Li, "The Relationship Between User Experience and Machine Learning," *SSRN Electron. J.*, no. 1, pp. 1-11, 2018, doi: 10.2139/ssrn.3173932.

[31] V. Betigiri, "AI based UI Development (AI-UI)," *Medium*, 2018. .

[32] K. Siau and W. Wang, "Building Trust in Artificial Intelligence, Machine Learning, and Robotics," *Cut. Bus. Technol.*, vol. 31, no. 2, pp. 47-53., 2018, [Online]. Available: https://www.cutter.com/sites/default/files/itjournal/2018/cbtj1802.pdf.

[33] K. Darlington, "AI systems dealing with human emotions: how the future will be like with emotional machines," *OpenMind BBVA*, 2018. https://www.bbvaopenmind.com/en/technology/artificial-intelligence/ai-systems-dealing-with-human-emotions/ (accessed Jan. 30, 2021).

[34] EC-Council, "HOW TO USE ARTIFICIAL INTELLIGENCE FOR THREAT INTELLIGENCE," 2020. .

[35] C. Gonzalez, "6 Threat Modeling Methodologies: Prioritize & Mitigate Threats," 2020. https://www.exabeam.com/information-security/threat-modeling/ (accessed Dec. 26, 2020).

[36] J. E. T. Akinsola, A. Kuyoro, M. A. Adeagbo, and A. A. Awoseyi, "Performance Evaluation of Software using Formal Methods," *Glob. J. Comput. Sci. Technol. C Softw. Data Eng.*, vol. 20, no. 1, 2020, [Online]. Available: https://computerresearch.org/index.php/computer/article/view/1930/1914.

[37] S. K. Mandava, "User Interfaces with Artificial Intelligence," *Artificial Intelligence, User Interface, User Experience*, 2020. .

[38] R. Strehlow, "Cyber Security Requires an Important Ingredient," *Strong UX*, 2018. .

[39] H. Jaye, "What Is A User Interface, And What Are The Elements That Comprise one ?," *UI Design*, 2019. .

[40] B. Ganapathy, "How Artificial Intelligence is transforming Human-Computer Interaction, and its implications for Design," *LinkedIn*, 2017. https://www.linkedin.com/pulse/how-artificial-intelligence-transforming-interaction-its-ganapathy/ (accessed Jan. 30, 2021).

[41] S. Kleber, "3 Ways AI Is Getting More Emotional," *Harvard Business Publishing*, 2018. https://hbr.org/2018/07/3-ways-ai-is-getting-more-emotional (accessed Jan. 30, 2021).

[42] M. Redondo, C. Bravo, and M. Ortega, "Intelligent User Interfaces: Past, Present and Future," *Eng. User Interface*, pp. 1-12, 2009, doi: 10.1007/978-1-84800-136-7_18.

[43] D. Sonntag, "Intelligent User Interfaces Design and Implementation," 2016.

[44] R. Tahir, "Analyzing the intelligence in user interfaces," in *SAI Intelligent Systems Conference*, 2015, pp. 674-680, doi: 10.1109/IntelliSys.2015.7361213.

[45] T. G. Gonçalves, K. M. De Oliveira, E. Grislin-Le Strugeon, C. Kolski, and G. H. Travassos, "A systematic literature review on intelligent user interfaces: Preliminary results," in *IHM 2019 - Annexes des Actes de la 31e Conference Francophone sur l'Interaction Homme-Machine*, 2019, pp. 1-8, doi: 10.1145/3366551.3370344.

[46] T. G. Gonçalves and A. R. Cavalcanti da Rocha, "Development process for intelligent user interfaces: An initial approach," 2019, doi: 10.1145/3364641.3364665.

[47] K. Höök, "Steps to take before intelligent user interfaces become real," *Interact. Comput.*, vol. 12, no. 4, pp. 409-426, 2000, doi: 10.1016/S0953-5438(99)00006-5.

[48] S. Shaikh, M. Ajmal, N. Ahmed, and F. Badar, "Comprehensive Understanding of Intelligent User Interfaces," *Int. J. Adv. Comput. Sci. Appl.*, vol. 8, no. 6, pp. 395-401, 2017, doi: 10.14569/ijacsa.2017.080652.

[49] H. Lieberman, "User interface goals, AI opportunities," *AI Mag.*, vol. 30, no. 4, pp. 16-22, 2009, doi: 10.1609/aimag.v30i4.2266.

[50] J. Dudley and P. Kristensson, "A Review of User Interface Design for Interactive Machine Learning," *ACM Trans. Interact. Intell. Syst.*, vol. 1, no. 1, pp. 1-37, 2018.

[51] C. Loitsch, "Designing accessible user interfaces for all by means of adaptive systems," no. May 1982, 2018.

[52] S. Duggirala, "10 Usability Heuristics with Examples," *prototypr. io*, 2016. .

[53] D. Khourshid, "Mind-Reading with Adaptive and Intelligent User Interfaces - YouTube," 2020. https://www.youtube.com/watch?v=adO2crvd3fc&t=805s (accessed Jan. 09, 2021).

[54] Fuselab Creative, "Intelligent Interface User Design. Why It Is Important to Understand The Purpose of IUI |," 2018. https://fuselabcreative.com/intelligent-interface-user-design-when-ai-ui-converge/ (accessed Jan. 09, 2021).

[55] Fuselab Creative, "Intelligent Interface Design–What to Expect in 2019 |," 2019. https://fuselabcreative.com/emerging-trends-in-intelligent-interface-design-what-to-expect-in-2019/ (accessed Jan. 09, 2021).

[56] O. D. Alao, J. V Joshua, and J. E. T. Akinsola, "Human Computer Interaction (HCI) and Smart Home Applications," *IUP J. Inf. Technol.*, vol. 15, no. 3, pp. 7-21, 2019, Accessed: Jan. 02, 2021. [Online]. Available: https://search.proquest.com/openview/70e74bf39099ec671c013b7bf9d9258a/1?pq-origsite=gscholar&cbl=2029987.

[57] H. W. Alomari, V. Ramasamy, J. D. Kiper, and G. Potvin, "A User Interface (UI) and User eXperience (UX) evaluation framework for cyberlearning environments in computer science and software engineering education," *Heliyon*, vol. 6, no. 5, May 2020, doi: 10.1016/j.heliyon.2020.e03917.

[58] UXTOOLS, "Compare Prototyping Tools | UXTools.co - Uxtools.co," 2020. https://uxtools.co/tools/prototyping/ (accessed Jan. 11, 2021).

[59] Keep It Usable, "24 Top UX Prototyping Tools with Downloadable Comparison Table - Learn UX," 2020. https://www.keepitusable.com/blog/top-ux-prototyping-tools/ (accessed Jan. 11, 2021).

[60] S. Ogunsola, "Introduction to Uizard as a Rapid Prototyping Tool | by Soliudeen Ogunsola | Prototypr," *Prototypr*, 2019. https://blog.prototypr.io/introduction-to-uizard-as-a-rapid-prototyping-tool-ab3b6bb8729e (accessed Jan. 11, 2021).

[61] B. Wilkins, "Sketching Interfaces – Airbnb Design," 2020. https://airbnb.design/sketching-interfaces/ (accessed Jan. 11, 2021).

[62] N. Shevchenko, "Threat Modeling: 12 Available Methods," 2018. https://insights.sei.cmu.edu/sei_blog/2018/12/threat-modeling-12-available-methods.html (accessed Nov. 05, 2020).

[63] J. Fruhlinger, "Threat modeling explained: A process for anticipating cyber attacks | CSO Online," *CSO*, 2020. https://www.csoonline.com/article/3537370/threat-modeling-explained-a-process-for-anticipating-cyber-attacks.html.

[64] F. Shull, "Cyber Threat Modeling: An Evaluation of Three Methods," 2016. https://insights.sei.cmu.edu/sei_blog/2016/11/cyber-threat-modeling-an-evaluation-of-three-methods.html (accessed Jan. 08, 2021).

[65] N. Shevchenko, B. R. Frye, and C. Woody, "THREAT MODELING: EVALUATION AND RECOMMENDATIONS," 2018. [Online]. Available: https://apps.dtic.mil/sti/pdfs/AD1083907.pdf.

[66] V. S. Moustakis and J. Herrmann, "Where do machine learning and human-computer interaction meet?," *Appl. Artif. Intell.*, vol. 11, no. 7-8, pp. 595-609, 1997, doi: 10.1080/088395197117948.

[67] S. Blitz, "3 Ways Predictive Analytics Can Boost Your

Cybersecurity," *Sisense Inc.*, 2017. https://www.sisense.com/blog/3-ways-predictive-analytics-can-boost-cybersecurity/ (accessed Jan. 08, 2021).

[68] N. Shevchenko, T. A. Chick, P. O'riordan, T. P. Scanlon, and C. Woody, "THREAT MODELING: A SUMMARY OF AVAILABLE METHODS," 2018. [Online]. Available: https://resources.sei.cmu.edu/asset_files/WhitePaper/2018_019_001_524597.pdf.

[69] J. E. T. Akinsola, A. S. Ogunbanwo, O. J. Okesola, I. J. Odun-Ayo, F. D. Ayegbusi, and A. A. Adebiyi, "Comparative Analysis of Software Development Life Cycle Models (SDLC)," vol. 1, pp. 310-322, 2020, doi: 10.1007/978-3-030-51965-0_27.

[70] N. van Esch and F. Heijnen, "Design principles for AI-enabled UI | Deloitte Digital," *Deloitte*, 2020. https://www2.deloitte.com/nl/nl/pages/customer-and-marketing/articles/design-principles-for-ai-enabled-ui.html (accessed Jan. 12, 2021).

[71] LIFARS, "What Is Threat Modeling in Cybersecurity? A Brief Introduction," 2020. https://lifars.com/2020/10/what-is-threat-modeling-in-cybersecurity/.

[72] O. Caspi, "Vulnerability Management: How to Think Like an Attacker," *AT&T Business*, 2021. https://cybersecurity.att.com/resource-center/white-papers/vulnerability-management-think-like-an-attacker (accessed Jan. 08, 2021).

[73] E. Segal, "AI Applications in Cybersecurity with Real-Life Examples," *AltexSoft*, 2020. https://www.altexsoft.com/blog/ai-cybersecurity/ (accessed Jan. 08, 2021).

[74] CodeSealer, "Importance of user Interface protection from cyber attacks - CodeSealer," *Codesealer blog*, 2018. https://codesealer.com/what-is-user-interface-protection/ (accessed Dec. 25, 2020).

[75] L. Christou, "You shouldn't fear an AI lie detector, unless, of course, you have something to hide," *verdict*, 2019. https://www.verdict.co.uk/ai-lie-detector/ (accessed Jan. 30, 2021).